NUMBER POWER

5 THE REAL WORLD OF ADULT MATH

GRAPHS, TABLES, SCHEDULES AND MAPS

ROBERT MITCHELL & DONALD PRICKEL

Consultants:
Michael Dean
Elizabeth Shaw
VTAE—Adult H.S. Program
Broward County, Florida

Project Editor:
Caren Van Slyke

Contemporary Books, Inc.
Chicago

Published by Contemporary Books, Inc.
180 North Michigan Avenue, Chicago, Illinois 60601
International Standard Book Number: 0-8092-5516-2

Published simultaneously in Canada by Beaverbooks, Ltd.
195 Allstate Parkway, Valleywood Business Park
Markham, Ontario L3R 4T8 Canada

Production Editors: Geraldine Lynch, Debby Eisel

Editorial Assistant: Ellen Frechette

Illustrations: Ophelia Chambliss-Jones

TABLE OF CONTENTS

USING NUMBER POWER

TO THE STUDENT

NUMBER POWER 5 introduces you to the skills needed to read a wide variety of graphs, schedules, charts and maps. These skills can be used on standardized tests, at school, on the job, or in daily living.

The first section of the book, BUILDING NUMBER POWER, introduces you to reading and interpreting these kinds of materials. Each chapter begins with a skills inventory that gives an opportunity to test the skills you already have. Each chapter ends with a final skills inventory so that you can check your progress.

In the second section of the book, USING NUMBER POWER, you can work with types of graphs, charts, schedules, and maps that you might encounter in daily living.

To get the most out of your work, do each problem carefully and check each answer to make sure you are working accurately. An answer key is provided at the back of the book.

GRAPHS
GRAPH SKILLS INVENTORY

The Graph Skills Inventory allows you to measure your skills in reading and interpreting graphs.

GRAPH A

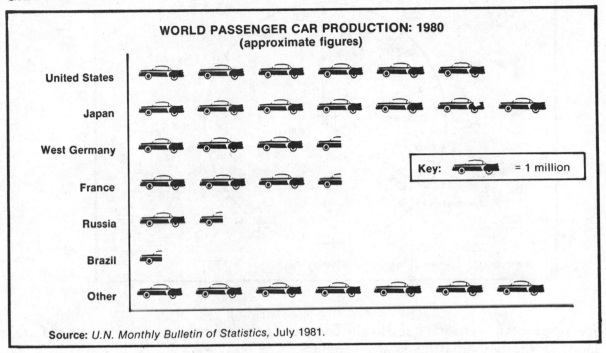

WORLD PASSENGER CAR PRODUCTION: 1980
(approximate figures)

United States

Japan

West Germany

France

Russia

Brazil

Other

Key: = 1 million

Source: *U.N. Monthly Bulletin of Statistics*, July 1981.

Directions: Answer each question by completing the sentence, answering true or false, or by choosing the best multiple-choice response.

1. The two countries that produced the greatest number of cars are the United States and _____.

2. The United States produced _____ cars in 1980.

3. Graph A represents all the passenger cars produced in the world in 1980.   True   False

4. Russia and Brazil produced approximately an equal number of cars in 1980.   True   False

5. The total production of cars by the six specific countries shown in Graph A was _____.
   a) $2\frac{1}{2}$ million
   b) $8\frac{1}{2}$ million
   c) 22 million
   d) 29 million
   e) 34 million

6. During 1980, Japan produced _____ as many passenger cars as West Germany.
   a) one-half
   b) twice
   c) one-fourth
   d) four times
   e) several times

GRAPH B

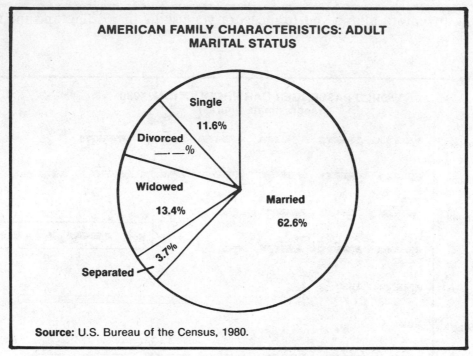

AMERICAN FAMILY CHARACTERISTICS: ADULT MARITAL STATUS

Single 11.6%

Divorced —.—%

Widowed 13.4%

Married 62.6%

3.7%

Separated

Source: U.S. Bureau of the Census, 1980.

Directions: Answer each question by completing the sentence, answering true or false, or by choosing the best multiple-choice response.

7. The categories of marital status shown are—married, separated, _____, divorced, and _____.

8. In 1980, the percent of the American population who were divorced was _____. (Fill this in on the graph.)

9. More adults were separated and divorced than were single.   True   False

10. In 1980, three-quarters of adult Americans were married.   True   False

11. The percent of adult Americans who were not single in 1980 was _____.

    a) 88.4%
    b) 11.6%
    c) 91.3%
    d) 50%
    e) 8.7%

12. In 1980, more than half of all adult Americans were _____.

    a) single or widowed
    b) divorced or widowed
    c) married
    d) separated or single
    e) widowed or separated

GRAPH C

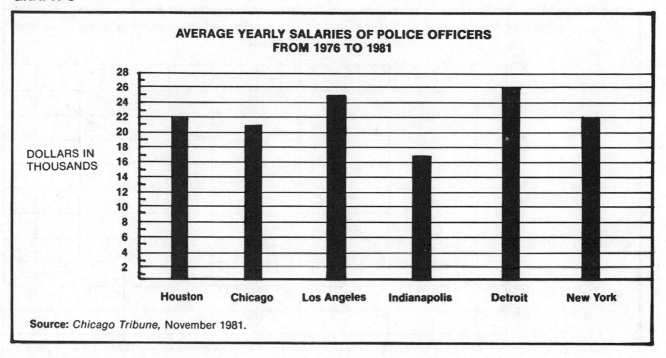

AVERAGE YEARLY SALARIES OF POLICE OFFICERS
FROM 1976 TO 1981

DOLLARS IN THOUSANDS

Houston   Chicago   Los Angeles   Indianapolis   Detroit   New York

Source: *Chicago Tribune*, November 1981.

Directions: Answer each question by completing the sentence, answering true or false, or by choosing the best multiple-choice response.

13. Graph C shows the average yearly salaries for _____.

14. The average yearly salary for police officers working in Chicago was _____.

15. Average yearly salaries are shown for New York and Denver.   True   False

16. Police officers in Detroit earned an average of $26,000 each year.   True   False

17. Police officers in Indianapolis earn _____ than officers in New York.
    a) $6,000 more
    b) $5,000 less
    c) $2,000 more
    d) $6,000 less
    e) $5,000 more

18. The average yearly salary of police officers in the six cities shown on Graph C is approximately _____.
    a) $17,000
    b) $22,000
    c) $26,000
    d) $44,000
    e) $133,000

GRAPH D

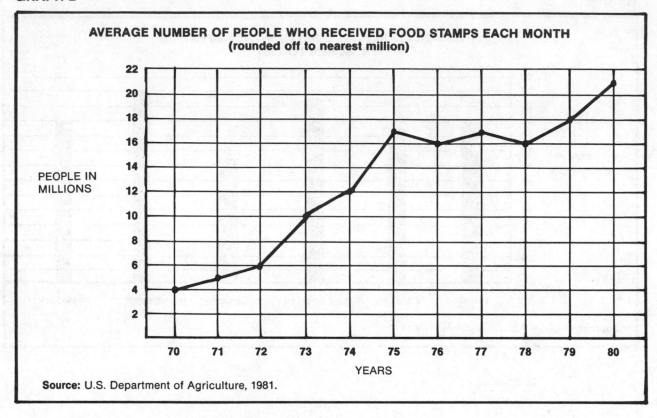

AVERAGE NUMBER OF PEOPLE WHO RECEIVED FOOD STAMPS EACH MONTH
(rounded off to nearest million)

PEOPLE IN MILLIONS

YEARS

Source: U.S. Department of Agriculture, 1981.

Directions: Answer each question by completing each sentence, answering true or false, or choosing the best multiple-choice response.

19.   Graph D shows the average number of people receiving
_____.

20.   In 1980, approximately _____ million people received food stamps monthly.

21.   In 1970, the average number of people receiving food stamps each month was 3 million.   True   False

22.   From 1979 to 1980, there was an increase in people receiving food stamps.   True   False

23.   From 1970 to 1975, there was an increase of _____ million people receiving food stamps.
a) 5
b) 9
c) 13
d) 17
e) 26

24.   The greatest increase in the number of food stamp recipients was during _____.
a) 1973
b) 1974
c) 1975
d) 1979
e) 1980

GRAPH SKILLS INVENTORY CHART

Use this inventory to see what you already know about graphs and what you need to work on. A passing score is 21 correct answers. Even if you have a passing score, circle the number of any problem that you miss, correct it, and turn to the practice page indicated.

Circle the Questions Missed	Type of Graph	Type of Question	Practice Page
3	Pictograph	Scanning	10
7	Circle		22
13, 15	Bar		34
19	Line		46
2	Pictograph	Reading	11
10, 12	Circle		23
14, 16	Bar		35
20, 21	Line		47
1, 4, 5, 6	Pictograph	Comprehension	11
8, 9, 11	Circle		23
17, 18	Bar		35
22, 23, 24	Line		47

WHAT ARE GRAPHS?

A graph is a pictorial display of information. Since it is drawn, rather than written, a graph makes it possible to get a quick impression of a great deal of data and to easily make comparisons and draw conclusions. Graphs are often used in government, business, and education and may appear in reports, newspapers, and magazines.

TYPES OF GRAPHS

In this workbook, you will study four main types of graphs.

Pictographs

A *pictograph* uses pictures or symbols to display information.

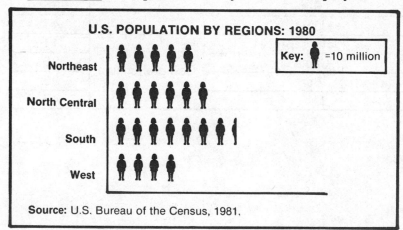

U.S. POPULATION BY REGIONS: 1980

Key: 👤 =10 million

Northeast
North Central
South
West

Source: U.S. Bureau of the Census, 1981.

A pictograph usually has a *key* to show the value of each symbol.

Pictographs are read by counting the symbols on a line of a graph and computing their value.

Circle Graphs

A *circle graph* uses parts of a circle to show information.

THE GOVERNMENT DOLLAR

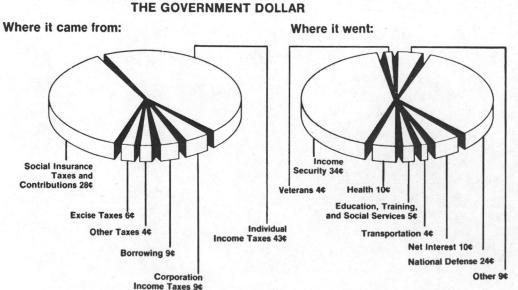

Where it came from:

Social Insurance
Taxes and
Contributions 28¢

Excise Taxes 6¢

Other Taxes 4¢

Borrowing 9¢

Corporation
Income Taxes 9¢

Individual
Income Taxes 43¢

Where it went:

Income
Security 34¢

Veterans 4¢   Health 10¢

Education, Training,
and Social Services 5¢

Transportation 4¢

Net Interest 10¢

National Defense 24¢

Other 9¢

Source: U.S. Office of Management and Budget, Fiscal Year 1981.

Circle graphs show values in each part of a divided circle. A part of a circle graph is called a *segment* or a *section*.

The segments of a circle add up to a whole or to 100% of the topic.

Bar Graphs

A *bar graph* uses thick bars to show information.

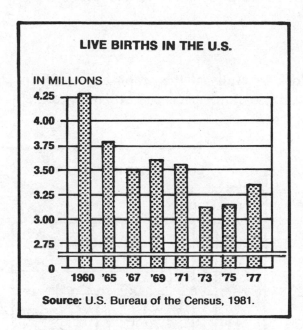

LIVE BIRTHS IN THE U.S.

IN MILLIONS

Source: U.S. Bureau of the Census, 1981.

Bar graphs are usually drawn in one of two different directions:

1) With the bars <u>running up and down</u>. The bars are placed at equal distances along the *horizontal axis* that runs across the bottom of the graph.

2) With the bars <u>running from left to right</u>. The bars are placed at equal distances along the *vertical axis* on the left side of the graph.

Bar graphs may also use a key to show additional information.

Sometimes, a graph may have a break in the vertical axis and an open space running across the graph. This means that some values have been left off to save space on the graph.

Line Graphs

A *line graph* is drawn with one or more thin lines that extend across the graph.

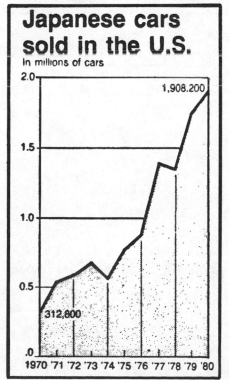

Japanese cars sold in the U.S.

In millions of cars

1,908.200

312,800

Source: Copyrighted, 1981, *Chicago Tribune.*

Used with permission.

Like the bar graph, a line graph is drawn using values along a horizontal and a vertical axis.

A line graph is most useful in showing trends and developments.

TYPES OF QUESTIONS

In this book, you are asked three types of questions. These questions will help you to find information and interpret graphs. Similar questions will be used in the sections on schedules, charts, and maps.

1) Scanning the Graph Questions

"Scanning the Graph" questions require you to look carefully at the graph and to fill in missing words to complete a sentence. To complete these statements, pay attention to the:

- title of the graph
- names of axes or sections (of a circle graph)
- information in the key, if used
- source of the graph's information, if given

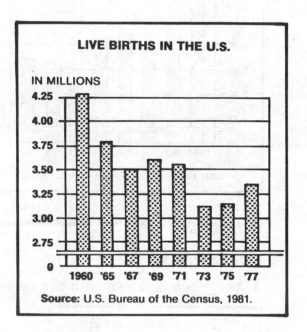

Example: This graph tells about _____ in the United States.

Answer: **live births** The title of the graph gives this information.

2) Reading the Graph Questions

"Reading the Graph Questions" require you to determine specific information from the graph. You should respond either true or false to the given statements.

Example: In 1967, there were $3\frac{1}{2}$ million live births in the United States.   True   False

Answer: **True.** Starting at the bottom of the graph (horizontal axis), find 1967. Then, look at the top of the 1967 bar and follow a line across to a point on the vertical axis to the left labeled 3.50 million live births (which is the same as $3\frac{1}{2}$ million).

3) Comprehension Questions

Comprehension questions require you to do computations, make inferences, draw conclusions, or make predictions based on the graph's information. You will circle the letter of one of five choices that best completes each sentence.

Example: The two years that had nearly the same number of births were ____.

a) 1955 and 1960
b) 1965 and 1967
c) 1971 and 1973
d) 1973 and 1975
e) 1977 and 1979

Answer: **d) 1973 and 1975.** Compare the heights of the bars. These bars are closer in height than any other two years.

READ CAREFULLY TO AVOID MISTAKES

Since interpretation is very important when answering questions, be sure to read a question carefully before choosing your response. Some questions may be tricky. The categories below give you some idea of possible problem areas.

1) Information is True but not Contained on Graph

Sample Question:

According to Graph A, Japan sells more cars to the United States than it does to any other country.   True   False

Answer: **False.** While this fact may or may not be true, the graph does not compare Japan's sales in the U.S. with its sales to other countries.

2) Misleading Words are Given in Answer Choices

Sample Question:

The statement that best describes Graph A is:

a) The percentage of Japanese cars sold in the U.S. is increasing.

b) The percentage of Americans driving Japanese cars is increasing.

c) The number of Japanese cars sold in the U.S. is increasing.

Answer: **c)** The word "percentage" is misleading in the choices "a" and "b." The graph does not show percentages; it shows only numbers of cars sold.

GRAPH A

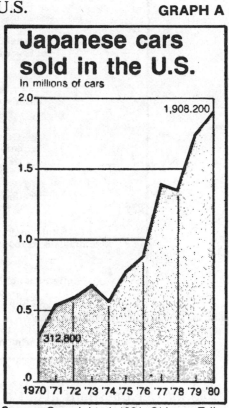

Japanese cars sold in the U.S.
In millions of cars

Source: Copyrighted, 1981, *Chicago Tribune.*
Used with permission.

3) Partial Symbols and Fractions

Sample Question:

In 1974, the number of Japanese cars sold in the U.S. was approximately:

a) 0.5 car
b) $\frac{1}{2}$ million cars
c) 500 cars

Answer: **b)** $\frac{1}{2}$ million cars. Often, responses given for a question may have the same value but be in a different form. The number of cars sold in 1974 is shown on the graph as 0.5, but if you read the small title over the vertical axis, you will see that this stands for 0.5 million cars. Since 0.5 million is not one of the given answers, read the choices carefully and determine that 0.5 million is the same as $\frac{1}{2}$ million.

PICTOGRAPHS

A *pictograph* gets its name from the small pictures it uses as symbols on the graph. Pictographs generally use a *key* to show the value of the pictures that are used as symbols. Parts of symbols are often used to represent a fractional amount of a quantity shown in the key.

Pictographs are often not as exact as other types of graphs, but they are the easiest to read.

GRAPH A

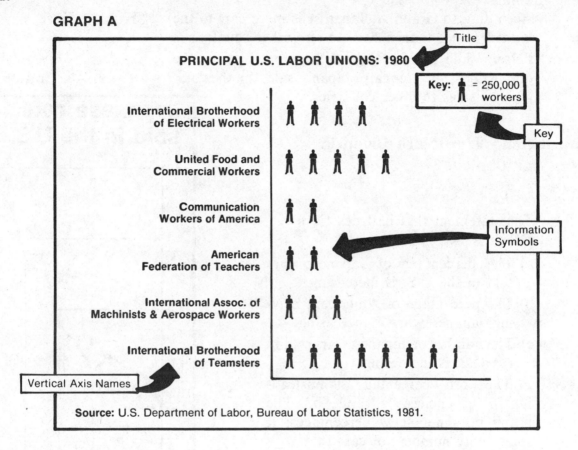

Source: U.S. Department of Labor, Bureau of Labor Statistics, 1981.

To answer questions about a pictograph, follow the sequence below.

SCANNING THE GRAPH

To scan a pictograph, find the graph title, the vertical axis names, and the key.

Sample Fill-In Question:

The Communication Workers of America is one of the principal _____.

Step 1. Find the name on the vertical axis, "Communication Workers of America."

Step 2. Look at the graph title: "Principal U.S. Labor Unions: 1980." This title gives the subject of the graph.

Answer: U.S. labor unions

READING THE GRAPH

To read a pictograph, count the number of symbols on a line. Then, multiply the number of symbols by the value of the symbol given in the key.

Sometimes, only a part of the symbol is shown. Look at the partial symbol carefully. Most often, a partial symbol will be ½ of the whole or sometimes ¼ or ¾. To find a value for a part of a symbol, find that fraction of the whole.

Sample True-Or-False Question:

(Circle your answer.)

The total membership of workers in the International   True   False
Brotherhood of Teamsters is 1,875,000.

Step 1. Find the name "International Brotherhood of Teamsters" on the vertical axis.

Step 2. Count the number of complete symbols: seven (7). Next, determine the fraction that the partial symbol represents: one-half symbol (½). There are 7½ symbols on that line.

Step 3. Compute the value of the symbols. (1) Multiply the whole numbers. (2) Multiply the fraction. (3) Add.
(1) $250,000 \times 7 = 1,750,000$
(2) $250,000 \times \frac{1}{2} = $ +125,000
(3) 1,875,000

Answer: True

COMPREHENSION QUESTIONS

Values pictured on the graph can be compared and conclusions can be drawn.

Sample Multiple-Choice Question:

(Circle the letter of the answer that best completes the sentence.)

The difference in membership between the American Federation of Teachers (AFT) and the International Brotherhood of Electrical Workers (IBEW) is _____.

a) 250,000 workers
b) 400,000 workers
c) 750,000 workers
d) 500,000 workers
e) 100,000 workers

Step 1. Find the name on the vertical axis, "American Federation of Teachers." Find this union's membership total by multiplying the number of symbols by the key's value. $2 \times 250,000 = 500,000$

Step 2. Find the name, "International Brotherhood of Electrical Workers." Find this union's membership by multiplying the number of symbols by the key's value. $4 \times 250,000 = 1,000,000$

Step 3. Subtract to find the difference between the memberships of the two unions.
IBEW = 1,000,000
AFT = −500,000
       500,000

Answer: d) 500,000 workers

PRACTICE GRAPH I

Graph I, read from left to right across the page, is an example of a horizontal pictograph. Graph I shows the hourly earnings of jobs in particular industrial and manufacturing groups during 1981.

GRAPH I

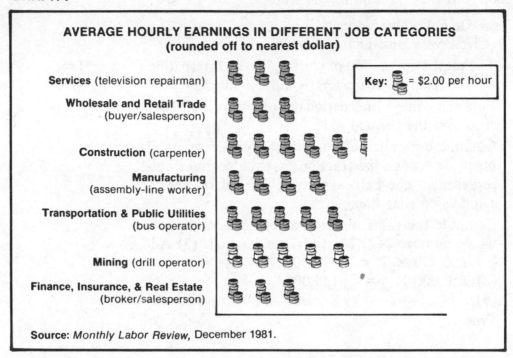

AVERAGE HOURLY EARNINGS IN DIFFERENT JOB CATEGORIES
(rounded off to nearest dollar)

Source: *Monthly Labor Review,* December 1981.

Scanning Graph I

Fill in each blank as indicated.

1. Find each of the following:
    a) Graph Title: _____

    b) Vertical Axis Names:

    _____,

    _____,

    _____,

    _____,

    _____,

    _____,

    _____.

2. Each 🪙 symbol is equal to _____ in earnings per hour.

3. The hourly pay for a carpenter is found by looking at the job category called _____.

Reading Graph I

Decide whether each statement is true or false and circle your answer.

4. The average hourly earnings of a construction worker is $14.00 per hour.　　　True　　False

5. A bus operator can expect to earn average hourly wages of approximately $10.00 per hour.　　　True　　False

6. Graph I shows that salespeople working in clothing stores usually earn an average of $8.00 per hour.　　　True　　False

Comprehension Questions

Circle the letter of the answer that best completes each sentence.

7. Hourly earnings in the wholesale and retail trades, finance, and service industries are generally _____ other occupations.

   a) higher than
   b) lower than
   c) the same as
   d) twice that of
   e) none of the above

8. The difference between the hourly wages of an assembly line worker and the earnings of a carpenter is _____.

   a) $1.50
   b) $2.00
   c) $3.00
   d) $4.00
   e) $5.00

9. The job category shown by Graph I to have the highest hourly earnings is _____.

   a) mining
   b) manufacturing
   c) transportation
   d) construction
   e) finances

10. The statement that best describes Graph I is:
    a) More physically difficult jobs pay higher hourly wages.
    b) Workers skilled in a trade or a craft make lower wages than workers skilled in management or supervision.
    c) Average hourly earnings for union workers range from $2.00 to $3.00 per hour higher than for non-union workers.
    d) Average hourly earnings for workers in manufacturing and industrial occupations can range from $6 to $11 per hour.
    e) Average hourly earnings for non-union workers are increasing at a rate faster than earnings for union workers.

PRACTICE GRAPH II

Graph II is a vertical pictograph. Symbols are shown in vertical columns. A vertical pictograph is often used to compare information about one item over a period of time. Graph II shows the change in the unemployment rate over several years.

GRAPH II

UNEMPLOYMENT RATES: SHOWN AS A PERCENT OF
CIVILIAN LABOR FORCE

Key: % = one (1) percent

	1969	1971	1973	1975	1977	1979	1981
				%			
				%			%
				%	%		%
		%		%	%	%	%
		%	%	%	%	%	%
	%	%	%	%	%	%	%
	%	%	%	%	%	%	%
	%	%	%	%	%	%	%
	%	%	%	%	%	%	%

Source: U.S. Department of Labor, Bureau of Labor Statistics, December 1981.

Scanning Graph II

Fill in each blank as indicated.

1. The information presented on Graph II was obtained from the source known as the _____
   _____.

2. The unemployment rate of the labor force is given from the year _____ to the year _____.

3. Graph II tells about the unemployment rates as a _____ of the civilian labor force.

Reading Graph II

Decide whether each sentence is true or false and circle your answer.

4. The highest unemployment rate shown occurred in   True   False
   1981.

5. In 1979, the unemployment rate was about 5½%.   True   False

6. According to Graph II, the lowest unemployment rate   True   False
occurred in the year 1973.

Comprehension Questions

Circle the letter of the answer that best completes each sentence.

7. The difference in the unemployment rate between 1971
and 1973 was _____.

a) 1%
b) 2%
c) 3%
d) 4%
e) 5%

8. The greatest unemployment rate gain occurred between
the years _____.

a) 1969 and 1971
b) 1971 and 1973
c) 1973 and 1975
d) 1975 and 1977
e) 1979 and 1981

9. From Graph II, you could assume that the greatest
percentage of the workforce was employed in the year
_____.

a) 1969
b) 1975
c) 1977
d) 1979
e) 1981

10. The statement that best describes Graph II is:
   a) Unemployment rates rose steadily from 1969 to
   1981.
   b) Unemployment rates rose steadily from 1973 to
   1981.
   c) Unemployment rates went up and down with a low
   point in 1969 and a high point in 1975.
   d) Unemployment rates follow a consistent pattern of
   increasing one year and then decreasing the next
   year.
   e) Unemployment rates were higher in 1981 than in any
   other year since 1969.

PRACTICE GRAPH III

For the purpose of making comparisons, some pictographs, such as Graph III below, are used to display more than one type of information.

GRAPH III

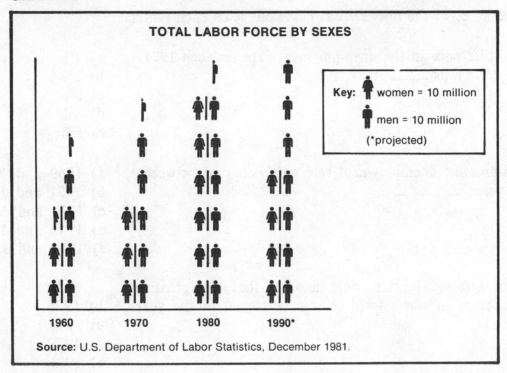

Scanning Graph III

Fill in each blank as indicated.

1. In Graph III, the asterisk (*) means that the information for 1990 is _____.

2. Each 👨 or 👩 on the graph represents _____ workers.
    (number)

3. Graph III is a labor force comparison between _____ and _____.

Reading Graph III

Decide whether each sentence is true or false, and circle your answer.

4. The smallest number of women was in the work force during the 1960's.   True   False

5. During the 1970's, women made up more than one-half of   True   False
    the <u>total</u> labor force.

6. During the 1960's, the <u>total</u> labor force was approximately   True   False
    90 million people.

Comprehension Questions

Circle the answer that best completes each sentence.

7. During the 1970's, the number of men in the work force   a) equal to
    was almost _____ the number of women.   b) twice
                                                                c) triple
                                                                d) four times
                                                                e) one-half

8. Between 1970 and 1980, the female labor force increased   a) 175 million
    by _____ .   b) 85 million
                                                              c) 30 million
                                                              d) 100 million
                                                              e) 40 million

9. The diagram that best illustrates a feature of Graph III is:

a)   b)   c)   d)   e)

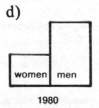
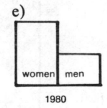

1980   1980   1980   1980   1980

10. The statement(s) that is (are) true for Graph III is (are):
    a) In comparison to the 1980's, the number of people in
       the total labor force is expected to decline in the
       1990's.
    b) Men have been and are expected to continue as the
       larger of the two groups of employed workers.
    c) By 1990, women are expected to make up a majority of
       the labor force.
    d) a and b
    e) a and c

APPLYING YOUR SKILLS: PICTOGRAPHS

When workers are unemployed, their families undergo many hardships. Some people may be able to find other work when they are fired or laid off. Others may not be so fortunate and must rely on unemployment compensation and public assistance for a while. One way to guard against becoming unemployed is to develop skills that will be in demand in the future.

JOB TRAINING—PROTECTION AGAINST UNEMPLOYMENT

Many people no longer plan to work for a lifetime at one type of job. They carefully examine developments in the job market so that they can acquire the appropriate training . . .

GRAPH A

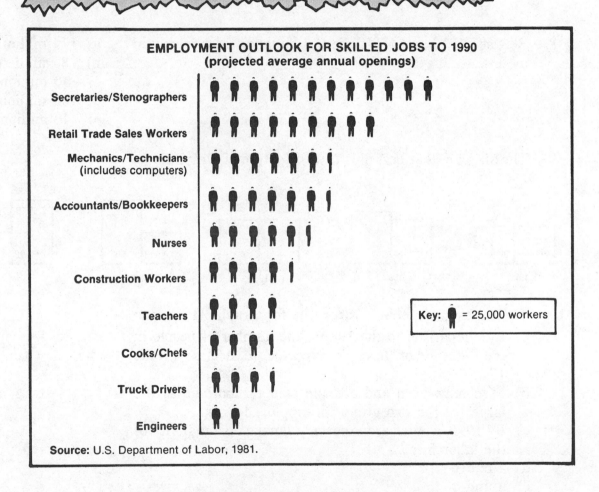

EMPLOYMENT OUTLOOK FOR SKILLED JOBS TO 1990
(projected average annual openings)

Secretaries/Stenographers

Retail Trade Sales Workers

Mechanics/Technicians
(includes computers)

Accountants/Bookkeepers

Nurses

Construction Workers

Teachers

Cooks/Chefs

Truck Drivers

Engineers

Key: ❘ = 25,000 workers

Source: U.S. Department of Labor, 1981.

Directions: Answer each question by completing the sentence, choosing true or false, or selecting the best multiple-choice response.

1. Graph A shows the employment outlook for _____ _____ to the year 1990.

2. The total number of job openings for secretaries/stenographers in each year to 1990 will be approximately _____.

3. The need for stenographers/secretaries is expected to be three times greater than the need for school teachers.   True   False

4. According to Graph A, the number of job openings for construction workers will be greater in 1990 than it is in 1983.   True   False

5. The information shown in Graph A would indicate that individuals having typing, sales, and computer programming skills would have better prospects for employment than those without.   True   False

6. The average number of yearly job openings for nurses during the 1980's will be approximately _____.

    a) 50 thousand
    b) 90 thousand
    c) 140 thousand
    d) 160 thousand
    e) 230 thousand

7. According to Graph A, the difference in job openings between cooks and _____ is approximately 50,000 per year.

    a) teachers
    b) nurses
    c) accountants
    d) secretaries
    e) sales workers

8. If the number of job openings in teaching increases 10% more than Graph A predicts, the total number of yearly job openings in teaching will then be _____.

    a) 110,000
    b) 100,000
    c) 50,000
    d) 10,000
    e) 200,000

9. The diagram that best represents a relationship shown on Graph A is:

a)	b)	c)	d)	e)
teachers / engineers	truck drivers / secretaries	teachers / cooks	sales / nurses	cooks / truck drivers

10. All of the conclusions below can be drawn from Graph A EXCEPT:
    a) To 1990, jobs for cooks and teachers will be less available than jobs for accountants.
    b) During the 1980's, secretarial and stenographic skills will provide a person with greater opportunities for employment than any other listed skill.
    c) To 1990, the number of yearly openings will continue to be greater in the technical trades than in the nursing profession.
    d) The number of job openings for computer technicians will begin to decline after 1990.
    e) Until 1990, the number of engineers needed in the work force is expected to be substantially less than the number of sales workers.

Many unemployed workers must rely on government support to meet family and everyday expenses. The government helps workers who temporarily lose their jobs by providing unemployment compensation.

UNEMPLOYMENT BENEFITS: HOW STATES COMPARE

When job openings become scarce and workers are laid off, states provide compensation to the unemployed. The amount a state gives to the unemployed differs depending on . . .

GRAPH B

A NINE-STATE COMPARISON OF MAXIMUM UNEMPLOYMENT BENEFITS PER WEEK: 1981

State	Benefits
Connecticut	$ $ $ $ $ $ $ $ $ $
Florida	$ $ $ $
Illinois	$ $ $ $ $ $ $
Massachusetts	$ $ $ $ $ $ $ $ $
Michigan	$ $ $ $ $ $ $
Ohio	$ $ $ $ $ $ $ $ $ $
Oregon	$ $ $ $ $ $
Texas	$ $ $ $ $
Washington	$ $ $ $ $ $ $

Key: $ = $20.00

Source: National Commission on Unemployment Compensation, 1981.

Directions: Answer each question by completing the sentence, choosing true or false, or selecting the best multiple-choice response.

1. Graph B represents a nine-state comparison of _____ per week.

2. The state shown on Graph B that pays the lowest unemployment benefits is _____.

3. Ohio pays more than twice as much compensation to an   True   False
   eligible unemployed worker as does Texas.

4. Washington and Oregon pay lower unemployment benefits than either Ohio or Massachusetts.   True   False

5. Graph B shows that out of nine states, Connecticut pays the lowest dollar amount in unemployment benefits to its workers.   True   False

6. The states that pay the highest weekly benefits to an unemployed person are _____.

    a) Massachusetts and Ohio
    b) Washington and Michigan
    c) Connecticut and Ohio
    d) Oregon and Illinois
    e) Connecticut and Florida

7. The maximum that a person in Massachusetts could receive during a five-week period is approximately _____.

    a) $180.00
    b) $900.00
    c) $1,050.00
    d) $1,275.00
    e) $1,400.00

8. If a person collected $320 in maximum unemployment benefits during a four week period, that person could be living in the state of _____.

    a) Connecticut
    b) Illinois
    c) Ohio
    d) Oregon
    e) Florida

9. From Graph B, you can tell the maximum that unemployed workers in Massachusetts receive is _____ more per week than unemployed workers in Texas.

    a) $180.00
    b) $90.00
    c) $60.00
    d) $30.00
    e) $120.00

10. The statement that best describes Graph B is:
    a) Unemployment benefits among the nine states range from $80 to $190 per week.
    b) According to Graph B, all the states pay at least $95 per week in unemployment benefits to eligible individuals.
    c) Average unemployment benefits paid in Ohio to eligible individuals is $180.
    d) States prefer that unemployment benefits be paid by the federal government.
    e) None of the above.

CIRCLE GRAPHS

A *circle graph* shows an entire quantity divided into various parts. Each part of the circle is called a *segment* and has its own name and value. In most cases, values on circle graphs consist of either parts of a dollar or of a fraction or a percent of a whole.

A circle graph is also called a pie graph.

Circle graphs are often used to illustrate budgets and expenses. Graph A shows the sources for each dollar that the federal government receives.

GRAPH A

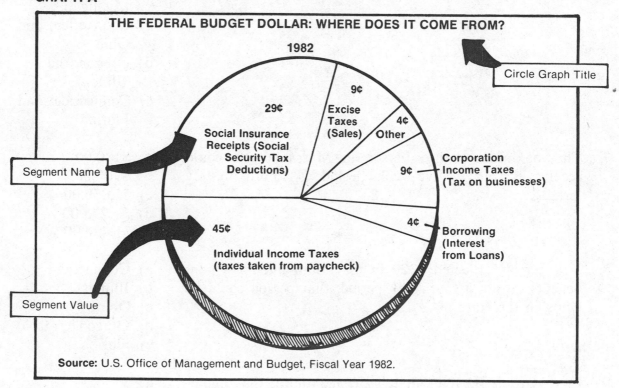

SCANNING THE GRAPH

To scan a circle graph, find and read the graph title and the segment names.

Sample Fill-In Question:
> Graph A shows that part of the federal budget dollar comes from individual income _____.

Step 1. Find and read the graph title, "The Federal Budget Dollar: Where Does It Come From?" The title is important, but does not give the answer.

Step 2. Read the segment names. Read clockwise until you find the segment name, "Individual Income Taxes." This is the answer.

Answer: taxes

READING THE GRAPH

To read a circle graph, locate the name and value of each segment. Each segment value can represent a fraction, percent, or number.

Sample True-Or-False Question:

(Circle your answer.)

From every dollar the government receives, 29¢   True   False
comes from social insurance receipts.

Step 1. Read clockwise on the graph until you find the segment named "Social Insurance Receipts."

Step 2. Near the segment name, read the dollar value. The value for social insurance receipts is 29¢.

Answer: True

COMPREHENSION QUESTIONS

By comparing segment names and segment values, you can draw conclusions from the graph.

Sample Multiple-Choice Question:

(Circle the letter of the answer that best completes the sentence.)

For each dollar, the difference between the amount   a) penny
that the government receives from excise taxes and   b) nickel
from borrowing is a _____.   c) dime

Step 1. Find the segment name and value "Excise Taxes—   d) quarter
9¢."   e) half-dollar

Step 2. Find the segment name and value "Borrowing—4¢."

Step 3. Find the difference between the two values by subtracting:

$$\begin{array}{r} 9¢ \\ -4¢ \\ \hline 5¢ \end{array} \text{ (a nickel.)}$$

Answer: b) nickel

PRACTICE GRAPH I

Graph I is a circle graph that shows where each dollar budgeted by the federal government is spent. Each pie-shaped segment represents a part of a dollar ($1.00).

GRAPH I

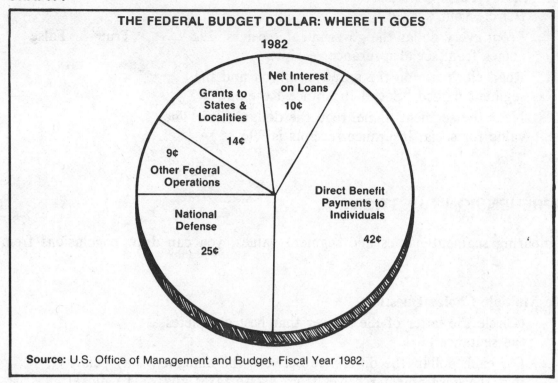

THE FEDERAL BUDGET DOLLAR: WHERE IT GOES

1982

Grants to States & Localities 14¢

Net Interest on Loans 10¢

9¢ Other Federal Operations

National Defense 25¢

Direct Benefit Payments to Individuals 42¢

Source: U.S. Office of Management and Budget, Fiscal Year 1982.

Scanning Graph I

Fill in each blank as indicated.

1. Find each of the following:
   a) This graph is about _____.
   b) The main categories of the federal budget are:

      _____

      _____

      _____

      _____

      _____

2. Direct benefit payments are made to _____.

3. The category to which Congress allocates money for missiles, tanks, and servicemen's pay is _____.

Reading Graph I

Decide whether each statement is true or false and circle your answer.

4. At least one-half of each federally budgeted dollar is used for direct benefit payments to individuals.   True   False

5. Of each dollar the government spends, 25% is spent on national defense.   True   False

6. The smallest amount spent in any one category is in grants to states and localities.   True   False

Comprehension Questions

Circle the letter of the answer that best completes each sentence.

7. From each dollar collected, the government spends _____ toward net interest on loans.

    a) 1%
    b) 5%
    c) 10%
    d) 25%
    e) 50%

8. The difference between money spent for national defense and money paid in benefits to individuals is _____ for each dollar.

    a) 48¢
    b) 73¢
    c) 23¢
    d) 9¢
    e) none of the above

9. During 1982, approximately _____ of the budget was spent on areas other than national defense.

    a) $\frac{1}{3}$
    b) $\frac{1}{4}$
    c) $\frac{2}{3}$
    d) $\frac{3}{4}$
    e) $\frac{4}{5}$

10. Graph I can be summarized by the statement:
    a) The federal dollar is equally divided among several categories.
    b) The federal government collects most of the money for its budget from personal income taxes.
    c) The two leading government expenditures are for direct benefit payments to individuals and national defense.
    d) The interest paid on the national debt increased 10% between 1981 and 1982.
    e) none of the above

PRACTICE GRAPH II

Graph II shows the 1982 budget allocations for Lane County, Oregon. From the graph, you can tell that the word "allocation" refers to the money that is to be divided and distributed to each of the various services in the county. Each segment of the circle represents a dollar amount and its percent of the total allocation.

GRAPH II

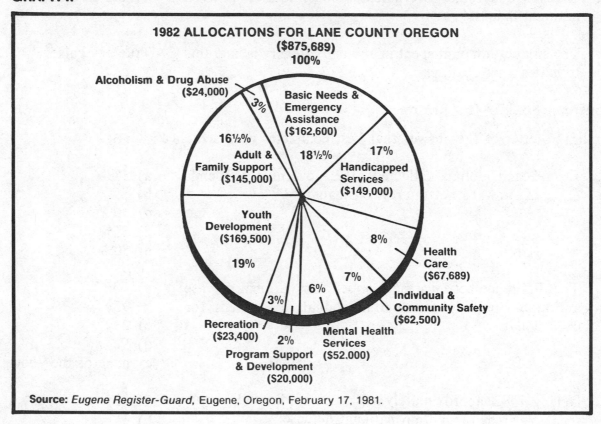

1982 ALLOCATIONS FOR LANE COUNTY OREGON
($875,689)
100%

Alcoholism & Drug Abuse ($24,000) — 3%

Basic Needs & Emergency Assistance ($162,600) — 18½%

Adult & Family Support ($145,000) — 16½%

Handicapped Services ($149,000) — 17%

Youth Development ($169,500) — 19%

Health Care ($67,689) — 8%

Individual & Community Safety ($62,500) — 7%

Mental Health Services ($52.000) — 6%

Program Support & Development ($20,000) — 2%

Recreation ($23,400) — 3%

Source: *Eugene Register-Guard*, Eugene, Oregon, February 17, 1981.

Scanning Graph II

Fill in each blank as indicated.

1. Graph II shows that 100% of the 1982 allocations for county services equals _____ dollars.

2. Money for a drug rehabilitation center would be allocated from the category titled _____.

3. The 1982 allocations are divided among _____
   (number)
   community programs.

Reading Graph II

Decide whether each sentence is true or false and circle your answer.

4. Health care will receive $16\frac{1}{2}$% of the 1982 allocations for county services.   True   False

5. Program support and development for 1982 will exceed   True   False
    $200,000.

6. Mental health services receive 6% of the 1982 budget   True   False
    allocations.

Comprehension Questions

Circle the letter of the answer that best completes each sentence.

7. Allocations for adult and family support exceed alloca-   a) $24\frac{1}{2}$
    tions for health care by _____ percent.   b) 8
    c) $16\frac{1}{2}$
    d) $8\frac{1}{2}$
    e) $14\frac{1}{2}$

8. If the percentage of health care allocations doubles in the   a) 8%
    next budget year, the new allocation for health care will be   b) 14%
    _____.   c) 16%
    d) 5%
    e) 17%

9. If the money budgeted for recreation for the following   a) $20,000
    year will be $2,000 less than that for program support in   b) $18,000
    1982, the dollar amount allocated for recreation will be   c) $21,400
    _____.   d) $23,400
    e) $25,400

10. All of the following statements can be inferred from
    Graph II EXCEPT:
    a) Meeting basic needs and giving emergency assistance
       are high budget priorities.
    b) A lower priority is given to program support and devel-
       opment than to any other category.
    c) A higher portion of budgeted dollars is given to
       youth development than to any other category.
    d) Lane County receives matching grants from the feder-
       al government.
    e) Lane County allocates money to assist the handi-
       capped.

PRACTICE GRAPH III

In some cases, more than one graph can be used to make comparisons. The following circle graphs show how the federal budget was estimated to change in 1982 and 1983 based on information known in 1981.

GRAPH III

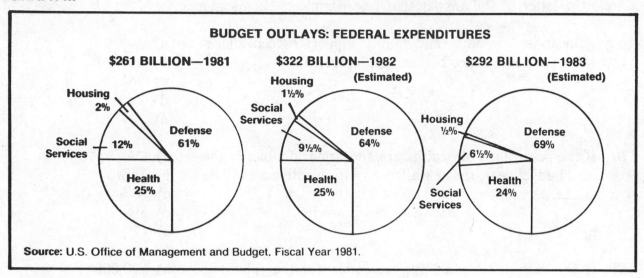

BUDGET OUTLAYS: FEDERAL EXPENDITURES

$261 BILLION—1981 $322 BILLION—1982 (Estimated) $292 BILLION—1983 (Estimated)

Source: U.S. Office of Management and Budget. Fiscal Year 1981.

Scanning Graph III

Fill in each blank as indicated.

1. The amount of federal expenditures shown for the years 1982 and 1983 are _____ figures.

2. The total budget for 1982 and 1983 are estimated to be $_____ billion and $_____ billion, respectively.

3. According to Graph III, the federal government's budget is divided into four main categories:

    a) _____

    b) _____

    c) _____

    d) _____

Reading Graph III

Decide whether each sentence is true or false and circle your answer.

4. The percentage of 1981 federal expenditures granted for housing was 2%.   True   False

5. The percent of the federal budget allocated for defense is expected to decrease over the three-year period.   True   False

6. The budget outlay for defense in 1983 is estimated to be 69%.   True   False

Comprehension Questions

Circle the answer that best completes each sentence.

7. From 1981 to 1983, the percent allocated for _____ will change the least.

    a) health
    b) housing
    c) social services
    d) defense
    e) all the above

8. The budget item that was projected to be reduced more than any other by the year 1983 is _____.

    a) defense
    b) housing
    c) education
    d) health
    e) all the above

9. The diagram that best shows the total budget outlays is:

a) 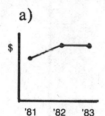 b) 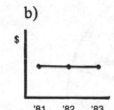 c) 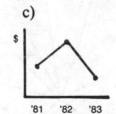 d) 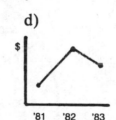 e)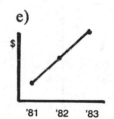

10. The statement that best reflects the information on the graphs is:
    a) For 1981 to 1983, there is a projected increase of total federal expenditures in all categories.
    b) For 1981 to 1983, there is a projected increase in the percentage of total federal expenditures going for defense.
    c) For 1982, total federal expenditures were projected to be less than they were in 1981 for both housing and health.
    d) Unemployment will increase in 1983 because of decreasing federal expenditures for social services.
    e) both b and d

APPLYING YOUR SKILLS: CIRCLE GRAPHS

When state and federal governments cut their budgets, the effects are felt by various agencies, organizations, and, most directly, by individual citizens. The following graphs reveal interesting information.

3 MILLION FACE LOSS OF FOOD STAMPS

Budget cutbacks are placing severe burdens on low-income families. The poor and unemployed depend, to some extent, on the government to help provide food, clothing, and other essentials.

GRAPH A

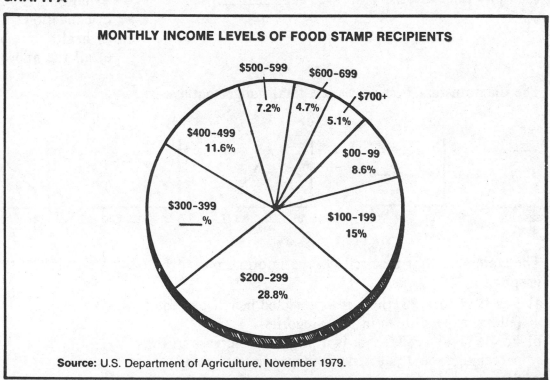

MONTHLY INCOME LEVELS OF FOOD STAMP RECIPIENTS

$500-599 7.2%
$600-699 4.7%
$700+ 5.1%
$400-499 11.6%
$00-99 8.6%
$300-399 ___%
$100-199 15%
$200-299 28.8%

Source: U.S. Department of Agriculture, November 1979.

Directions: Answer each question by completing the sentence, choosing true or false, or by selecting the best multiple-choice response.

1. Graph A shows the percentage of households receiving food stamps that are earning a specific _____ _____.

2. The percent of food stamp recipients in the $300 to $399 category must be _____. (Fill this amount in on the graph.)

3. Households earning in excess of $700 each month do not   True   False
    qualify for food stamps.

4. Graph A shows that 15% of households receiving food   True   False
    stamps earn between $100–$199 per month.

5. The largest percentage of households that receive food   True   False
    stamps earn from $300–$399 per month.

6. 11.6% of all households receiving food stamps are within   a) $100–$199
    the monthly income range of _____.   b) $200–$299
    c) $300–$399
    d) $400–$499
    e) $500–$599

7. _____ percent of all households that receive food   a) 83%
    stamps earn less than $400 each month.   b) 71.4%
    c) 19%
    d) 47.8%
    e) none of the above

8. Approximately twice as many households earning   a) $100–$199
    $200–$299 each month receive food stamps as households   b) $300–$399
    earning _____ each month.   c) $400–$499
    d) $500–$599
    e) $600–$699

9. The diagram that best illustrates the relationship of percentages on Graph A is:

a)   b)   c)   d)   e)

| $100–199 | $700+ | | $700+ | $100–199 | | $700+ | $100–199 | | $00–99 | $700+ | | $00–99 | $300–399 |

10. Graph A is best summarized as follows:
    a) The largest percentage of households that receive food
       stamps earn monthly incomes between $200–$299.
    b) Households with incomes above $700 each month are
       no longer eligible for food stamps.
    c) Household income levels are an important factor in
       determining the amount of food stamps received.
    d) both a and c
    e) both b and c

MILITARY ENLISTMENTS ON THE RISE
Due to poor prospects in the civilian job market, some military experts are forecasting increased enlistment in the armed services. Critics charge that high unemployment is encouraging large numbers of minorities to enlist, but that they are still seriously underrepresented in the officer corps.

GRAPH B

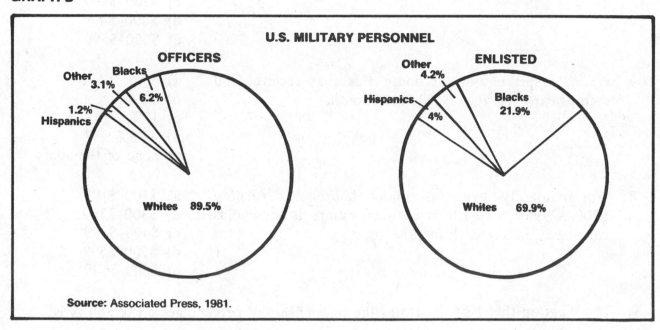

U.S. MILITARY PERSONNEL

OFFICERS
Other 3.1%
Blacks 6.2%
1.2% Hispanics
Whites 89.5%

ENLISTED
Other 4.2%
Hispanics 4%
Blacks 21.9%
Whites 69.9%

Source: Associated Press, 1981.

Directions: Answer each question by completing the sentence, choosing true or false, or selecting the best multiple-choice response.

1. The graphs above divide United States military personnel into the two general categories of _____ and _____.

2. An appropriate title for the two graphs above could be:
   a) Percent of Minorities in Military Service
   b) Number of Officers and Enlisted Military Personnel
   c) Military Personnel: Comparison by Racial Composition
   d) Ratio of Enlisted and Drafted Personnel
   e) Military Personnel and Civilian Population

3. Graph B shows that there are more officers than enlisted personnel in the United States military services.   True   False

4. There are about five times as many black officers as Hispanic officers.   True   False

5. From the graphs, you can infer that more enlisted personnel are becoming officers during the 1980's than in the 1970's.   True   False

6. The majority of military personnel consists of _____.
a) blacks
b) whites
c) Hispanics
d) other
e) none of the above

7. There are about _____ times the percent of Hispanic enlistees as Hispanic officers.
a) 2
b) 4
c) 5
d) 10
e) 20

8. Out of 100 military officers, Graph B would indicate that about _____ are white.
a) 12
b) 30
c) 10
d) 90
e) 60

9. The diagram that compares the percent of enlisted personnel to officers is:

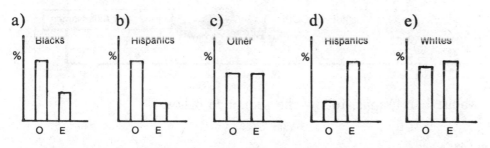

a) Blacks   b) Hispanics   c) Other   d) Hispanics   e) Whites

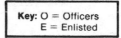

Key: O = Officers
E = Enlisted

10. All the statements below can be concluded from Graph B EXCEPT:
a) There is some representation of minorities in the officers' corps.
b) In the 1980's, the percent of minority officers is smaller than the percent of minority enlisted personnel.
c) Whites comprise the largest percentage of both the enlisted and officer categories.
d) More individuals are enlisting in the military in 1981 than in the past.
e) Slightly more than 10% of military officers are minorities.

BAR GRAPHS

Bar graphs get their name from the thick bars with which they are drawn. They are generally more complicated to read than pictographs or circle graphs.

Bar graphs contain labeled points (usually names or numbers) along the *horizontal* and *vertical axes*. The values represented by an individual bar, called an *information bar*, are determined by both its height (or length) and by its position along an axis. Follow the steps outlined below to learn how to accurately read bar graphs.

GRAPH A

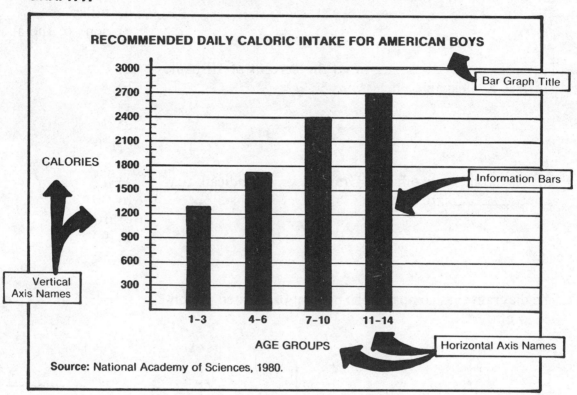

To answer questions about bar graphs, follow the sequence below.

SCANNING THE GRAPH

To scan a bar graph, find the graph title and the names of the axes.

Sample Fill-In Question:

Graph A shows the recommended daily calorie intake for American boys whose ages range from 1 to _____ years.

Step 1. Find the horizontal axis name "Age Groups."

Step 2. Scan from left to right, reading the labeled points at the bottom of each information bar. The bar farthest to the right represents the oldest age group, 11–14.

Answer: 14

READING THE GRAPH

To read a bar graph, find the values represented by the information bar. Read these values as labeled points along the horizontal and vertical axes.

Sample True-Or-False Question:

(Circle your answer.)

The amount of daily calories recommended for a boy 12 years of age is 2,700.

True   False

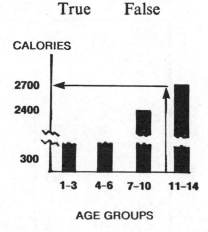

Step 1. Find the horizontal axis name, "11–14" years old. This axis name includes those boys whose ages are 11,12,13, and 14. Age 12 is included in this group.

Step 2. Move your eyes from the bottom to the top of this information bar. After doing this, move across the graph to the left until you reach the vertical axis.

Step 3. Read the vertical axis name and number, "Calories, 2,700." The number of calories for a boy 12 years old is 2,700.

Answer: True

COMPREHENSION QUESTIONS

Comprehension questions involve comparing values represented by several information bars and then drawing conclusions. Also, bar graphs are useful in showing trends. A trend is a pattern that can be seen from the information contained on the graph. From a trend, it is often possible to make predictions about future occurrences.

Sample Multiple-Choice Question:

(Circle the letter of the answer that best completes the sentence.)

As boys get older, the number of calories consumed daily should _____.

a) stay the same
b) increase
c) decrease
d) level off
e) none of the above

Step 1. Find the horizontal axis names that refer to the specific age groups. Scan from the bottom to the top of each bar and identify the number of calories recommended for each group.

Step 2. Compare the amount of calories between the ages of boys from very young to older. For example,

ages: 1–3 = 1,300 Calories
4–6 = 1,700 Calories
7–10 = 2,400 Calories
11–14 = 2,700 Calories

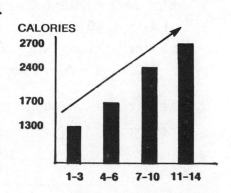

Step 3. Draw a conclusion from the information in Step 2: As boys get older, the number of calories they require increases.

Answer: b) increase

PRACTICE GRAPH I

Graph I shows a vertical bar graph with bars drawn up from the horizontal axis. Also, notice the symbol ⌇ on the vertical axis line. Occasionally, this symbol is used to show that values have been omitted from the axis in order to save space. On the graph below, pounds from 1 to 135 have been omitted.

GRAPH I

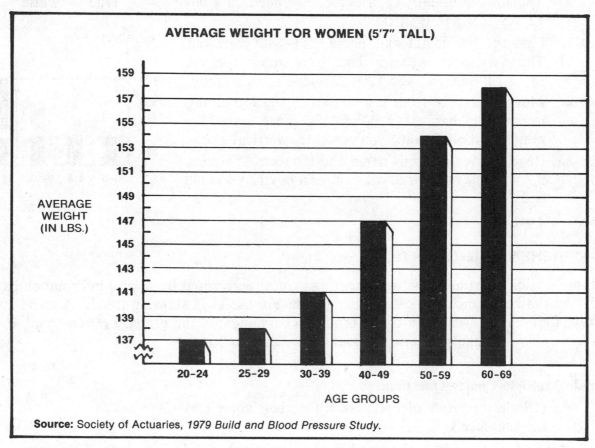

Scanning Graph I

Fill in each blank as indicated.

1. Find each of the following.
   a) Graph Title: _____
   b) Vertical Axis Title: _____
   c) Horizontal Axis Title: _____

2. The youngest women represented are from _____ to _____ years old.

3. The greatest average weight shown on the graph for women 5'7" tall is _____ pounds.

Reading Graph I

Decide whether each statement is true or false and circle your answer.

4. The average weight for a 30-year-old woman is 141   True   False
    pounds.

5. A 152-pound woman in the age group 30—39 years is   True   False
    below the average weight.

6. The highest average weight occurs for women in the   True   False
    50—59 year age group.

Comprehension Questions

Circle the letter of the answer that best completes each sentence.

7. On the average, as women get older, the trend is for their
    weight to _____ .

    a) decrease
    b) stay the same
    c) increase
    d) increase until 30
    e) none of the above

8. The difference in average weight between a 69-year-old
    woman and a 29-year-old woman is _____ pounds.

    a) 12
    b) 10
    c) 20
    d) 18
    e) 35

9. From Graph I, you can tell that the greatest gain in weight
    occurs from age group _____ to _____ .

    a) 20-24, 25-29
    b) 25-29, 30-39
    c) 30-39, 40-49
    d) 40-49, 50-59
    e) 50-59, 60-69

10. Graph I tells about the _____ .
    a) average weight and height for 5 foot 7 inch tall adults
    b) average weights for all women between the ages of 20
       and 69 years
    c) average weight at specific ages for women 5 feet 7
       inches tall
    d) a and b only
    e) b and c only

PRACTICE GRAPH II

Graph II is a horizontal bar graph. The information bars are drawn across the graph from left to right. Although they are not as common as vertical bar graphs, it is useful to know how to read horizontal bar graphs.

GRAPH II

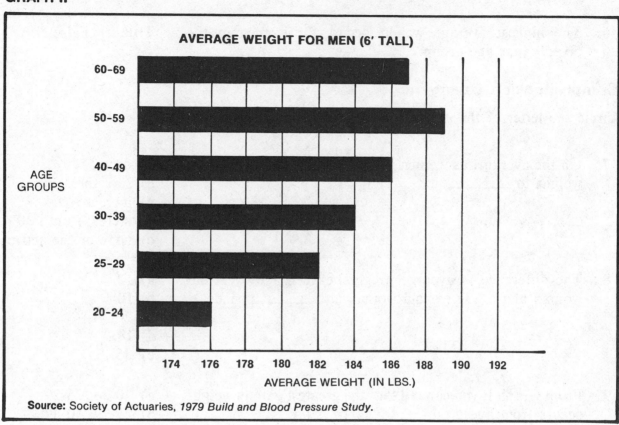

Scanning Graph II

Fill in each blank as indicated.

1. Graph II represents the _____ for men 6 feet tall.

2. Graph II gives information that concerns men who are _____ tall.

3. The total age span represented by Graph II starts at age _____ and extends to age _____.

Reading Graph II

Decide whether each sentence is true or false and circle your answer.

4. The average weight for a 6-foot-tall, 28-year-old man is   True   False
186 pounds.

5. The graph shows that the lowest average weight for all   True   False
men occurs in the 20 to 24 year age group.

6. A 6-foot-tall man who weighs 188 pounds is most likely to   True   False
be in the 50 to 59 year age group.

Comprehension Questions

Circle the letter of the answer that best completes each sentence.

7. On the average, a man's weight can be expected to in-   a) 45
crease until age _____.   b) 49
                                                              c) 39
                                                              d) 63
                                                              e) 59

8. The greatest average weight increase shown on the graph   a) 20-24 to 25-29
occurs from the ages of _____.   b) 25-29 to 30-39
                                                              c) 30-39 to 40-49
                                                              d) 40-49 to 50-59
                                                              e) 50-59 to 60-69

9. The weight trend for men 6 feet tall, between the ages of
20 and 69 is best shown by:

a) 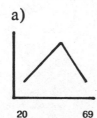   b) 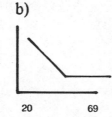   c) 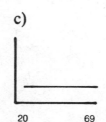   d) 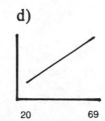   e)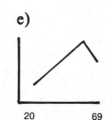

20   69   20   69   20   69   20   69   20   69

10. The statement that best describes the information on
Graph II is:
a) Men tend to gain weight as their age increases.
b) On the average, men 6 feet tall gain weight up to age
59, and tend to lose weight from age 60 on.
c) Taller men gain more weight in their later years than
shorter men.
d) Men tend to gain weight during retirement years.
e) Weight gain is a problem for most people.

PRACTICE GRAPH III

Graph III is a double bar graph that contains two kinds of information on the same graph. It shows the average weight for both men and women who are 5 feet 5 inches tall.

In each age group, a bar representing women's weight stands next to a bar representing men's weight. The bars are identified by a key in the upper right hand corner of the graph. Graphs containing more than one bar are used to make comparisons.

GRAPH III

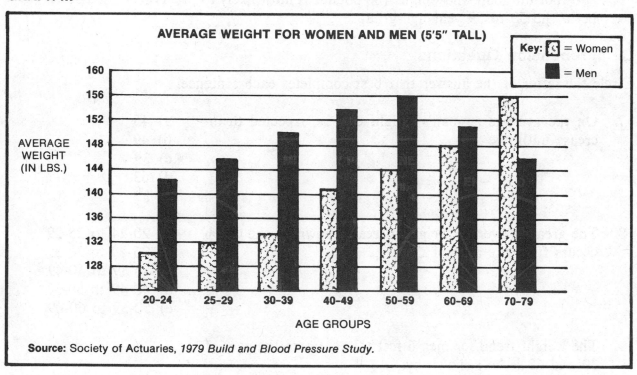

Source: Society of Actuaries, *1979 Build and Blood Pressure Study.*

Scanning Graph III

Fill in each blank as indicated.

1. In Graph III, ▮ represents the average weight for ____
_____, and ▨ represents the average weight
for _____.

2. The youngest age group represented on Graph III is from
age _____ to _____, and the oldest age group is from
_____ to _____.

3. The highest weight that could be represented on this graph is

_____.

Reading Graph III

Decide whether each sentence is true or false and circle your answer.

4. At age 53, a man is likely to weigh about 12 pounds more than a woman of the same age.   True   False

5. When both are in their seventies, men are usually heavier than women.   True   False

6. The ages at which the average weights of men and women are the closest to being equal are between 60 and 69 years.   True   False

Comprehension Questions

Circle the letter of the answer that best completes each sentence.

7. From 30 to 39 years, women weigh about _____ pounds less than men.
   a) 4
   b) 8
   c) 12
   d) 16
   e) 20

8. The trend on Graph III indicates that as a woman gets older, she is expected to _____ weight.
   a) gain
   b) lose
   c) gain and then lose
   d) lose and then gain
   e) none of the above

9. Women who are 5'5" tall and between the ages of 70–79 are the closest in weight to men 5'5" tall between the ages of _____.
   a) 20–24
   b) 25–29
   c) 40–49
   d) 50–59
   e) 60–69

10. The statement that best describes the information on Graph III is:
    a) In the age span shown, the pattern of weight gain for men is the same as the pattern of weight gain for women.
    b) Through their fifties, taller men and women are heavier than shorter men and women.
    c) Both men's and women's weight are affected by retirement.
    d) Both men and women tend to gain weight through their fifties, but men, unlike women, tend to lose weight in later years.
    e) None of the above.

APPLYING YOUR SKILLS: BAR GRAPHS

Americans are becoming more concerned with the quality of life and with their health. Weight and disease control, nutrition, and physical fitness are popular topics. Additionally, our society is placing great importance on improving health care services while controlling costs.

MEDICAL COSTS CONTINUE TO RISE

The costs of medical care continue to soar. Health care officials announced today that third party payments, including those by private insurance companies, government, and industry, are the highest ever, and that the . . .

GRAPH A

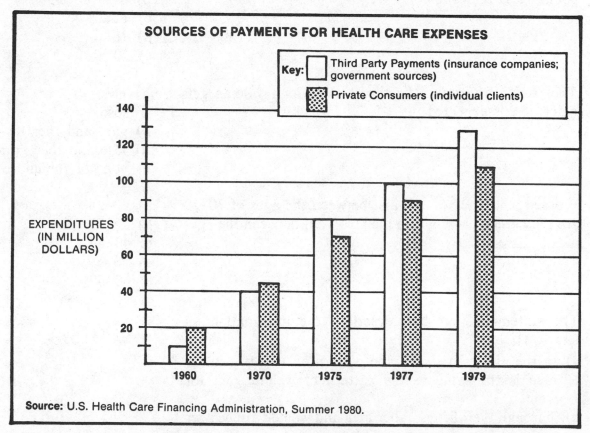

SOURCES OF PAYMENTS FOR HEALTH CARE EXPENSES

Key: ☐ Third Party Payments (insurance companies; government sources)

▨ Private Consumers (individual clients)

EXPENDITURES (IN MILLION DOLLARS)

Source: U.S. Health Care Financing Administration, Summer 1980.

Directions: Complete each sentence by filling in the blank, answering true or false, or selecting the best multiple-choice response.

1. The expenses of health care are generally paid by either _____ or _____.

2. In the year _____, third party payments began to be greater than private consumer payments.

3. Private consumer payments more than doubled from 1960 to 1970.   True   False

4. From 1960 to 1979, third party payments increased faster than private consumer payments.   True   False

5. The greatest yearly difference between private consumer and third party payments was in 1979.   True   False

6. The greatest increase in third party payments occurred from _____.

    a) 1975 to 1977
    b) 1960 to 1970
    c) 1977 to 1979
    d) 1970 to 1975
    e) none of the above

7. The total health care expenditures paid by both private consumers and third parties in 1979 was _____.

    a) $400 million
    b) $240 million
    c) $150 million
    d) $100 million
    e) none of the above

8. From 1977 to 1979, payments for health care (both private and third party) rose by _____ million.

    a) 30
    b) 40
    c) 50
    d) 60
    e) 70

9. According to Graph A, insurance companies and government payments began to account for one-half or more of the payments for health care expenses in _____.

    a) 1960
    b) 1970
    c) 1975
    d) 1977
    e) 1979

10. The statement that best describes the information on Graph A is:
    a) Third party payments tend to increase over the years while private consumer payments tend to decrease.
    b) Before 1975, private consumers paid the greater portion of health care costs.
    c) Third party payments should continue to be greater than private consumer payments in the future.
    d) Both a and b
    e) Both b and c

RAISE DRIVING AGE TO 18??

According to some public health researchers, accidents are the fourth most likely cause of death, following heart attacks, cancer, and strokes. . . . Raising the legal age for driving to age 18 would save at least 2,000 lives a year.

GRAPH B

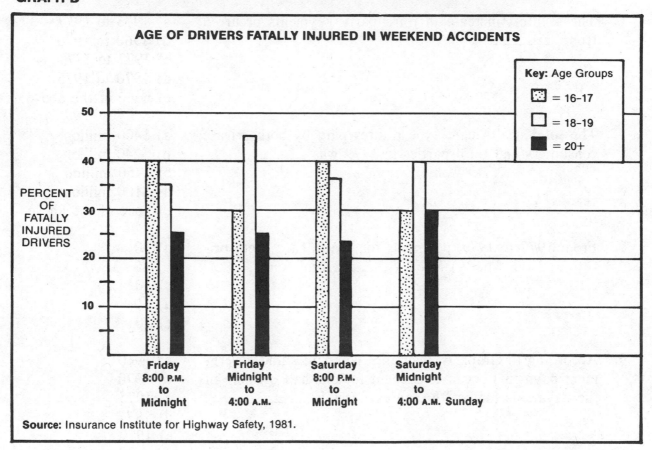

AGE OF DRIVERS FATALLY INJURED IN WEEKEND ACCIDENTS

Source: Insurance Institute for Highway Safety, 1981.

Directions: Answer each question by completing the sentence, choosing true or false, or selecting the best multiple-choice response.

1. Graph B represents drivers who are _____ in weekend car accidents.

2. Teenagers, 16 and 17 years of age, account for _____ percent of driver deaths on Saturday nights between 8:00 p.m. and midnight.

3. The same percent of 16 and 17-year-old drivers are fatally injured from 8:00 p.m. to midnight Friday and from 8:00 p.m. to midnight Saturday.   True   False

4. After midnight Saturday, more 16 and 17-year-old drivers have accidents than do drivers in the other two age groups.   True   False

5. 16- and 17-year-old drivers have the same percentage of fatal accidents as 20+ year-old drivers on _____.

a) Friday 8 p.m. to midnight
b) Friday midnight to 4:00 a.m.
c) Saturday 8 p.m. to midnight
d) Saturday midnight to 4:00 a.m.
e) none of the above

6. The highest percentage of fatally injured drivers, 20 years and older, is _____ and occurs on Saturday, midnight to 4:00 a.m.

a) 55%
b) 30%
c) 15%
d) 25%
e) 32%

7. The time period when the highest percentage of fatally injured drivers is in the 20-year-old and over driving group is _____.

a) Friday 8 p.m. to midnight
b) Friday midnight to 4:00 a.m.
c) Saturday 8 p.m. to midnight
d) Saturday midnight to 4:00 a.m.
e) none of the above

8. Of the groups listed, which has an average accident rate of 35% for the entire weekend? _____.

a) 16–17 year olds
b) 18–19 year olds
c) 20 + year olds
d) teenage men
e) impossible to tell from graph

9. The statement that best describes Graph B is:
a) Fatally injured drivers are most often 16 to 17 years old for every time period during the weekend.
b) Teenagers are more likely to have fatal accidents than adults.
c) On Friday night between 8:00 p.m. and midnight, 85% of all drivers who are fatally injured are 18 to 19 years old.
d) Of the three age groups, drivers who are 20 and older are the least likely to be fatally injured in a weekend car accident.
e) both b and d

LINE GRAPHS

Line graphs are especially useful in showing trends and developments over a period of time. A line graph can be drawn with either curved or straight lines that extend across the graph in a horizontal direction.

Graph A shows how consumer prices changed over a period of months.

GRAPH A

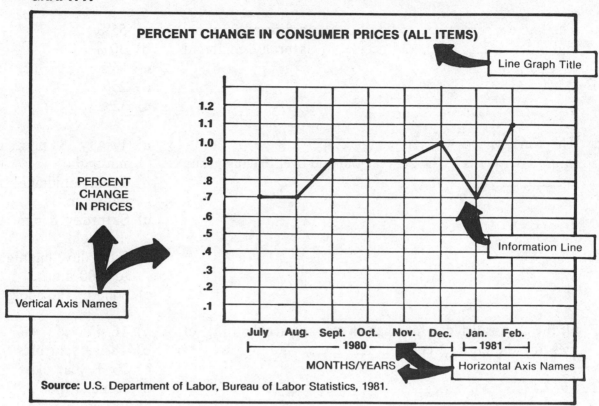

To answer questions about a line graph, follow the steps below.

SCANNING THE GRAPH

To scan a line graph, find the graph title, the axes names, and the labeled points along each axis.

Sample Fill-In Question:

Graph A shows the percent change in consumer prices for _____ months.

Step 1. Find the horizontal axis title: Months/Years.

Step 2. Count the number of months listed along the horizontal axis.

Answer: eight.

READING THE GRAPH

To read a line graph, find the information line, and read the labeled points along the horizontal and vertical axis.

Sample True-Or-False Question:

(Circle your answer.)

In December 1980, the change in consumer prices was 1.1%.

True   False

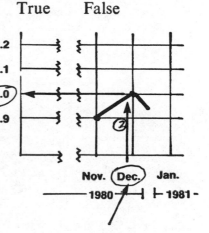

Step 1. Find December 1980, on the horizontal axis. From the bottom of the graph, move directly upward to the information line.

Step 2. From this point on the information line, move to the left to the labeled point on the vertical axis.

Step 3. Read the labeled point on the vertical axis as "1.0%."

Answer: False. In December 1980, consumer prices rose by 1.0%.

COMPREHENSION QUESTIONS

Inferences and predictions can be made by comparing values represented on the information line.

Sample Multiple-Choice Question:

(Circle the letter of the answer that best completes the sentence.)

The three consecutive months that had the same percent change in prices were _____.

a) June, July, February
b) July, August, January
c) September, October, November
d) October, November, December
e) August, September, October

Step 1. Scan across the information line from left to right. Identify the three consecutive points on the information line that are at the same height.

Step 2. For each of the three points on the information line, move directly downward to the horizontal axis name.

Step 3. Read the months directly below the designated points on the information line.

Answer: c) September, October, November

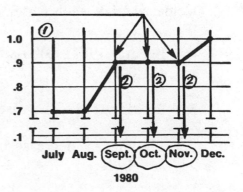

PRACTICE GRAPH I

Practice Graph I uses a single line to show the rise and fall of sugar prices.

GRAPH I

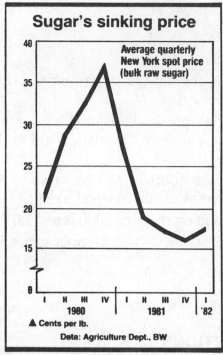

Sugar's sinking price

Average quarterly
New York spot price
(bulk raw sugar)

▲ Cents per lb.

Data: Agriculture Dept., BW

1. On the graph above, Roman numerals I, II, III and IV represent _____.

2. The graph measures the price of sugar in _____ per _____.

3. Each year is divided into _____ time periods.

Reading Graph I

Decide whether each statement is true or false and circle your answer.

4. In early 1981, 1,000 pounds of raw sugar would have cost $21,000.   True   False

5. By early 1982, the cost of sugar had begun to rise again.   True   False

6. The biggest decrease in the price of sugar took place in late 1981.   True   False

Comprehension Questions

Circle the letter of the answer that best completes each sentence.

7. The difference between the cost of a pound of sugar in the first quarter of 1980 and the first quarter of 1982 was about _____.

   a) $4.00
   b) $.04
   c) $38.00
   d) $.38
   e) $.17

8. The difference between the highest and the lowest raw sugar prices shown on the graph is about _____.

   a) $0.38
   b) $0.20
   c) $0.17
   d) $0.10
   e) $0.05

9. From the beginning of 1981 to the beginning of 1982, sugar prices _____.

   a) increased steadily
   b) decreased steadily
   c) decreased and then increased
   d) increased and then decreased
   e) stayed the same

10. From the information on the graph, you could infer that:
    a) Sugar workers went on strike in 1981.
    b) The price of manufactured sugar products rose in the early 1980's.
    c) The price of raw sugar will rise through the 1980's.
    d) Sugar prices reflect the rate of inflation.
    e) The rise in sugar prices has been a major cause of inflation and recession.

PRACTICE GRAPH II

To compare different types of information, a line graph uses more than one line. To prevent confusion, the lines are often drawn differently. A key is often used to indicate the meanings of the different lines.

On some line graphs, the vertical axis line may appear to be broken and some values skipped. This is done to omit values that are not important and to save space. On Graph II, the values between .10 and .40 have been omitted.

GRAPH II

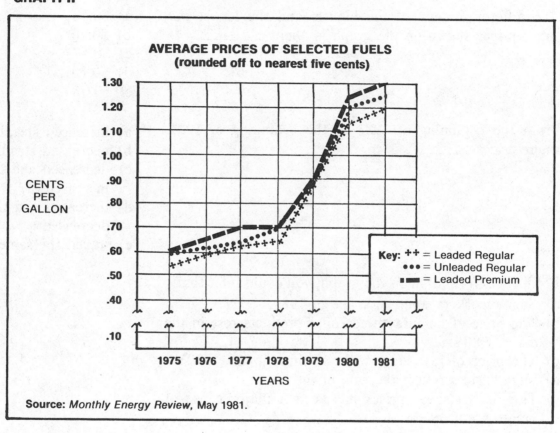

Source: *Monthly Energy Review,* May 1981.

Scanning Graph II

Fill in each blank as indicated.

1. On Graph II, the line represents _____ _____ fuel.

2. The average prices of gasoline are measured in _____ per _____.

3. Graph II shows gasoline prices for _____ years.

Reading Graph II

Decide whether each statement is true or false and circle your answer.

4. In 1975, leaded premium gasoline cost an average of $.55   True   False
per gallon.

5. The price of unleaded gasoline rose approximately $.25   True   False
per gallon from 1975 to 1978.

6. In 1978, 10 gallons of leaded premium gas would have   True   False
cost $7.00.

Comprehension Questions

Circle the letter of the answer that best completes each sentence.

7. Generally, the highest priced type of gasoline has been
_____.

    a) unleaded regular
    b) leaded premium
    c) leaded regular
    d) imported
    e) all are the same

8. Since 1975, the costs of all three types of gasoline have
_____.

    a) stayed the same
    b) increased slowly
    c) decreased
    d) more than doubled
    e) more than tripled

9. In the year _____, leaded and unleaded fuels cost the
same.

    a) 1975
    b) 1976
    c) 1977
    d) 1978
    e) 1979

10. A prediction that can be made from Graph II is:
    a) In a few years, leaded premium will no longer be the
       most costly type of gasoline.
    b) By the late 1980's, most automobiles will be using
       leaded premium gas.
    c) In the mid 1980's, the cost of all types of gasoline will
       decline.
    d) In the future, lower gas consumption will force prices
       up.
    e) None of the above.

PRACTICE GRAPH III

Graph III shows the consumer price indexes (cpi) for three categories of living expenses. Consumer price indexes are used by government and business to show the relative value of items in today's economy. Most consumer price indexes are based on what $100 would buy at 1967 prices. For example, if clothing has a cpi of 140, it means that it would cost $140 today to buy goods that cost $100 in 1967.

GRAPH III

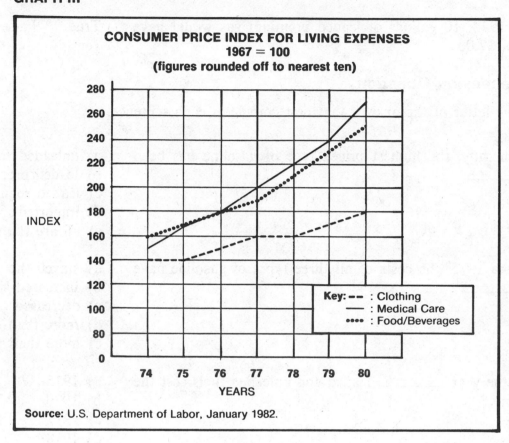

Scanning Graph III

Fill in each blank as indicated.

1. Graph III represents the consumer price indexes for _____, _____, and _____.

2. The consumer price index used in Graph III is based on costs in the year _____.

3. According to the key, ——— stands for _____.

Reading Graph III

Decide whether each statement is true or false and circle your answer.

4. Medical care has a consumer price index of 270 for 1980.   True   False

5. The consumer price index for food/beverages shows an   True   False
   increase since 1974.

6. The consumer price index for clothing doubled from 1975   True   False
   to 1980.

Comprehension Questions

Circle the letter of the answer that best completes each sentence.

7. The cpi is increasing most rapidly in the area of _____.
   a) clothing
   b) medical care
   c) food/beverages
   d) insurance
   e) none of the above

8. The total increase in the consumer price index for medical
   care from 1974 to 1980 was _____.
   a) 100
   b) 120
   c) 140
   d) 160
   e) 180

9. The years in which the consumer price index for
   food/beverages and medical care were the same occurred
   in _____.
   a) 1980 and 1981
   b) 1979 and 1980
   c) 1978 and 1979
   d) 1976 and 1977
   e) 1975 and 1976

10. The statement that best describes Graph III is:
    a) Medical care and food/beverage expenses rose at a
       faster rate than clothing expenses from 1974 to 1980.
    b) Between 1974 and 1980, clothing prices increased and
       medical care costs stayed the same.
    c) The three cpi's shown are increasing at about the same
       rate.
    d) The consumer price index for food/beverage and medi-
       cal care was twice as high in 1980 as it was in 1976.
    e) Inflation is the number one cause of the rise in the
       consumer price index.

APPLYING YOUR SKILLS: LINE GRAPHS

The cost of raising a child has risen greatly during the last several years and this trend is expected to continue in both the 1980's and the 1990's.

COSTS OF RAISING CHILDREN ON THE INCREASE

In this era of inflation, costs for health care, food, clothing, shelter, and education are expected to continue to increase. Experts have projected that by the year 2000, it will cost nearly $150,000 to raise a child. These figures are very . . .

GRAPH A

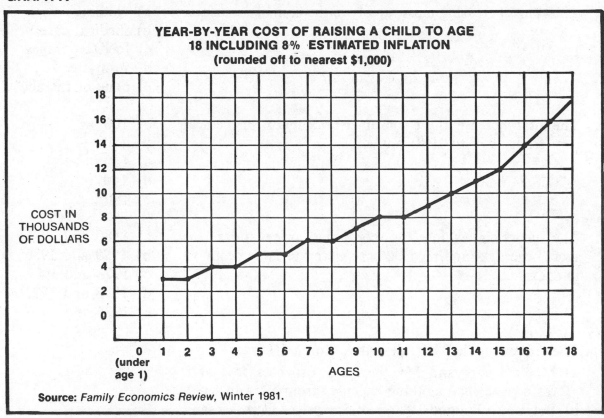

YEAR-BY-YEAR COST OF RAISING A CHILD TO AGE 18 INCLUDING 8% ESTIMATED INFLATION
(rounded off to nearest $1,000)

COST IN THOUSANDS OF DOLLARS

AGES

Source: *Family Economics Review*, Winter 1981.

Directions: Answer each question by completing the sentence, choosing true or false, or selecting the best multiple-choice response.

1. Graph A represents the cost of raising a child to age

——————.

2. The projected costs of raising a child shown on Graph A includes an estimated inflation rate of ————.

3. Overall, the yearly cost of raising a child is expected to increase as the child gets older.   True   False

4. The cost of raising two children from birth to age 5 is expected to be $22,000. (*Hint:* For example, the cost of raising a child between its second and third birthday is $4,000.)   True   False

5. Graph A shows that the yearly cost of supporting a 14 year old is $11,000.   True   False

6. The period during which the cost of raising a child increases most rapidly is from the ages of _____.
   a) 1 to 5
   b) 5 to 9
   c) 9 to 13
   d) 13 to 17
   e) 0 to 4

7. If 10% of the yearly cost of raising a child is for clothing, the cost of clothing a child at age 13 will be _____.
   a) $900
   b) $1,000
   c) $90
   d) $100
   e) $1,400

8. According to the graph, the total cost of raising a child to age 17 is projected to be approximately _____.
   a) $231,000
   b) $110,000
   c) $340,000
   d) $131,000
   e) $251,000

9. The cost of raising a child from the ages of 11 to 17 is _____ more than what it cost to raise the child from birth to age 10.
   a) $45,000
   b) $131,000
   c) $16,000
   d) $12,000
   e) $29,000

10. All the following statements can be concluded from Graph A EXCEPT:
   a) By the time a child is 17 years old, it is expected that he or she could cost $17,000 to feed, clothe, and educate.
   b) Overall, as a child gets older, the cost of raising the child gets higher.
   c) For the first two years, the cost of raising a child is relatively constant.
   d) The cost of raising a child in the first five years does not increase as rapidly as it does from 13 to 17.
   e) Inflation accounts for some of the increased costs of raising a child.

SAVE MONEY: WEATHERIZE YOUR HOME

Personal incomes are not increasing at the same rate as utility bills. Therefore, many families are cutting costs by taking steps to insulate and weatherize. The possible savings in weatherizing a home can . . .

GRAPH B

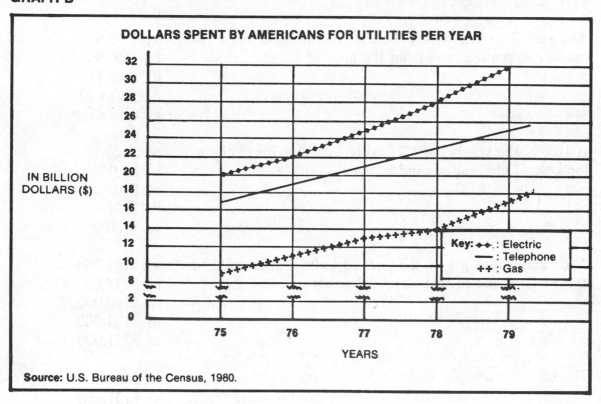

DOLLARS SPENT BY AMERICANS FOR UTILITIES PER YEAR

IN BILLION DOLLARS ($)

Key: ◆◆ : Electric
— : Telephone
++ : Gas

YEARS

Source: U.S. Bureau of the Census, 1980.

Directions: Answer each question by completing the sentence, choosing true or false, or by selecting the best multiple-choice response.

1. Graph B compares the money spent by Americans for
   _____ , _____ , and
   _____ .

2. According to Graph B, Americans spend more money on
   _____ than on gas or _____
   _____ .

3. In 1975, the total money spent for the three utilities was   True   False
   approximately $32 billion.

4. In 1976, Americans spent approximately $22 billion for telephone services.   True   False

5. From Graph B, you could conclude that heating costs are rising.   True   False

6. The greatest increase in dollars spent for electricity was from _____.

a) 1974 to 1975
b) 1975 to 1976
c) 1976 to 1977
d) 1977 to 1978
e) 1978 to 1979

7. In 1977, _____ more was spent for electricity than for gas.

a) $10 billion
b) $12 billion
c) $14 billion
d) $16 billion
e) $18 billion

8. In 1978, the total dollar amount spent for electricity was two times the amount spent for _____.

a) gas
b) telephone
c) both gas and telephone
d) all utilities
e) none of the above

9. The diagram that most accurately describes Graph B is:

a)   b)   c)   d)   e)

10. A conclusion that can be drawn from Graph B is:

a) American consumers are using more utilities every year.
b) American consumers are spending a larger part of their income for utilities every year.
c) Prices for utilities have increased more than prices in any other consumer area.
d) The cost of personal telephone bills has kept up with the rate of inflation.
e) Less money is spent by American consumers for gas than for other utilities.

FINAL GRAPH SKILLS INVENTORY

Do all the following problems. Work accurately, but do not use outside help. After completing the inventory, check your answers with the key on page 163. Your score will tell you how well you understand these graphs.

GRAPH A

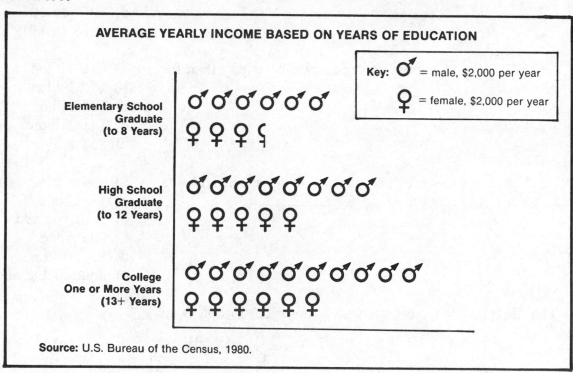

Directions: Answer each question by completing the sentence, answering true or false, or by selecting the best multiple-choice response.

1. Graph A shows income earned based on the amount of
    _____ a person has received.

2. From Graph A, you could infer that education helps to
    increase a person's chance for increased _____.

3. Men with 13+ years of education make an average salary   True   False
    of $20,000 per year.

4. Women elementary school graduates earn an average of   True   False
    $8,000 per year.

5. A female high school graduate earns about _____ than a male high school graduate.

    a) $6,000 less
    b) $3,000 more
    c) $6,000 more
    d) $3,000 less
    e) the same amount

6. On the average, the monthly income of a male high school graduate is about _____ more than a male who has graduated from elementary school.

    a) $600
    b) $300
    c) $700
    d) $100
    e) $500

7. On the average, a woman with one or more years of college education will earn _____ more per year than a woman with an elementary school education.

    a) $2,000
    b) $5,000
    c) $1,000
    d) $8,000
    e) $10,000

8. The statement(s) that best describe(s) Graph A is:
    a) The amount of education individuals receive has little effect on their income opportunities.
    b) The amount of education individuals receive tends to improve their income opportunities.
    c) Even with an equivalent educational level, men tend to earn more money than women.
    d) a and c
    e) b and c

GRAPH B

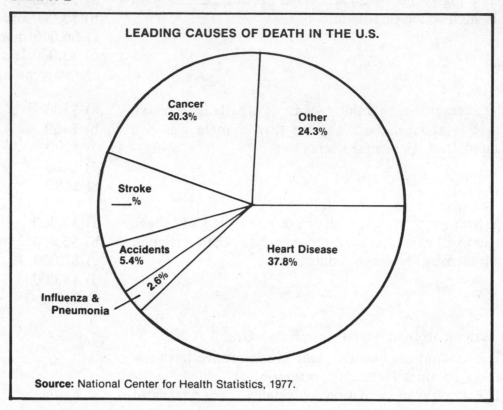

LEADING CAUSES OF DEATH IN THE U.S.

Cancer 20.3%

Other 24.3%

Stroke ___%

Accidents 5.4%

2.6%

Influenza & Pneumonia

Heart Disease 37.8%

Source: National Center for Health Statistics, 1977.

Directions. Answer each question by completing the sentence, answering true or false, or by selecting the best multiple-choice response.

9. Graph B, pictured above, represents the leading causes of _____ in the _____.

10. The percentage of deaths in the United States due to strokes is _____. (Fill this in on the graph.)

11. Heart disease and accidents are the two leading causes of death in the United States.   True   False

12. Each year, heart disease accounts for slightly more than one-third of the deaths in the United States.   True   False

13. The number of deaths due to accidents is _____ less than deaths due to strokes.

    a) 4.2%
    b) 3.6%
    c) 1.8%
    d) 17.8%
    e) 13.5%

14. Together, heart disease and _____ account yearly for slightly more than one-half of the deaths in the United States.

    a) heart attacks
    b) strokes
    c) cancer
    d) pneumonia
    e) other

15. A diagram that correctly shows the relationship between two causes of death is:

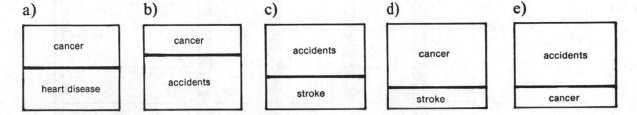

a) cancer / heart disease
b) cancer / accidents
c) accidents / stroke
d) cancer / stroke
e) accidents / cancer

16. All the statements can be concluded from Graph B EXCEPT:

    a) Cancer and heart disease cause more than 50% of the deaths in the United States.

    b) Fewer people die from car accidents each year in the United States than from strokes.

    c) The leading cause of death in the United States is heart disease.

    d) Graph B shows five leading causes of death in the United States.

    e) Research funding for heart disease exceeds the funding for all other fatal illnesses.

GRAPH C

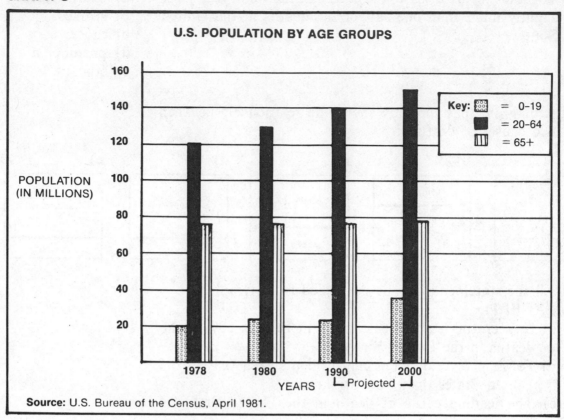

U.S. POPULATION BY AGE GROUPS

Key: = 0–19
= 20–64
= 65+

POPULATION (IN MILLIONS)

YEARS   Projected

Source: U.S. Bureau of the Census, April 1981.

Directions: Answer each question by completing the sentence, answering true or false, or by selecting the best multiple-choice response.

17. Graph C, pictured above, represents the population of the _____, and for each year shown, the population is divided into _____ age groups.

18. In 1980, the total population in the United States was _____.

19. Graph C shows the world population over a 22-year period.   True   False

20. By the year 2000, it is estimated that there will be about 150 million people who are 20 to 64 years old.   True   False

21. Between 1978 and 1980, the total increase in population in the United States was approximately _____ million.
a) 45
b) 2
c) 15
d) 30
e) 65

22. Between 1980 and 2000, it is estimated that the _____ age group will show the greatest gain in total number of people.

a) 0—19
b) 20—64
c) 65+
d) 0—12
e) 30—40

23. Between 1978 and 1990, it is estimated that the number of people in the 65+ age group will _____.

a) increase rapidly
b) decrease rapidly
c) remain almost constant
d) increase until 1990
e) none of the above

24. A conclusion that can be drawn from Graph C is:
a) The life expectancy of Americans is increasing.
b) Since 1978, older people are retiring at an earlier age than before 1978.
c) The total number of people in each age group shown is expected to be greater in the year 2000 than in the year 1978.
d) There were more teenagers in 1980 than the projections show for 1990.
e) both a and c

GRAPH D

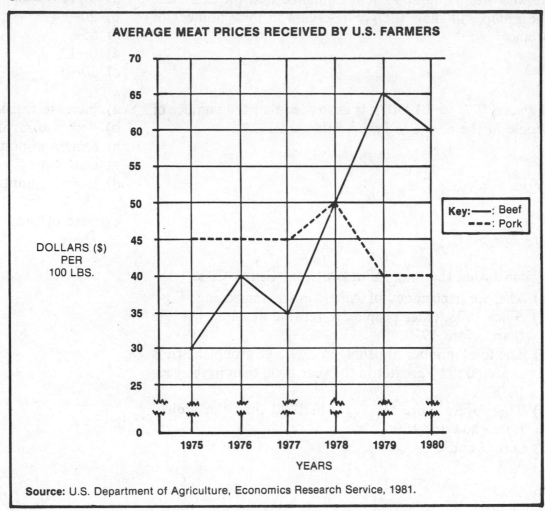

AVERAGE MEAT PRICES RECEIVED BY U.S. FARMERS

DOLLARS ($) PER 100 LBS.

YEARS

Key: —— : Beef
     --- : Pork

Source: U.S. Department of Agriculture, Economics Research Service, 1981.

Directions: Answer each question by completing the sentence, answering true or false, or by selecting the best multiple-choice response.

25. Graph D shows the average prices received by U.S. farmers for _____ and _____.

26. The highest price paid for beef occurred in the year _____ and was equal to $_____ per hundred pounds.

27. The prices appearing in Graph D represent the number of dollars for every 10 pounds of meat.   True   False

28. Prices for pork rose steadily from the years 1977 to 1979.   True   False

29. Farmers received the same price for both pork and beef during the year _____.

    a) 1975
    b) 1977
    c) 1978
    d) 1979
    e) 1980

30. During 1980, a farmer could earn approximately _____ more for 100 pounds of beef than for 100 pounds of pork.

    a) $50
    b) $40
    c) $30
    d) $20
    e) $10

31. The diagram that best represents information from Graph D is:

Key: P = Pork
B = Beef

a)
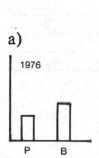

b)
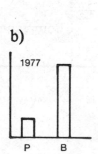

c)
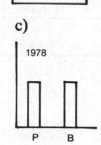

d)
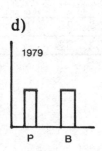

e)
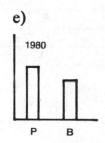

32. A conclusion that can be drawn from Graph D is:

a) Meat prices will decline during the 1980's.

b) In 1980, farmers could expect to earn more income from beef than from pork.

c) Pork and beef prices will rise and fall with the consumer price index.

d) The price of pork received by farmers can be expected to rise above the price of beef during the next several years.

e) The prices a consumer pays for beef and pork decreases as the average income received by U.S. farmers decreases.

FINAL GRAPH SKILLS INVENTORY CHART

Circle the number of any problem that you missed and be sure to review the appropriate page. A passing score is 27 correct answers. If you miss more than 5 questions, you should review this chapter.

Circle the Questions Missed	Type of Graph	Type of Question	Practice Page
1	Pictograph	Scanning	10
9	Circle		22
17, 19	Bar		34
25, 27	Line		46
3, 4	Pictograph	Reading	11
12	Circle		23
20	Bar		35
26	Line		47
2, 5, 6, 7, 8	Pictograph	Comprehension	11
10, 11, 13, 14, 15, 16	Circle		23
18, 21, 22, 23, 24	Bar		35
28, 29, 30, 31, 32	Line		47

SCHEDULES AND CHARTS SKILLS INVENTORY

This inventory allows you to measure your skills in reading and interpreting schedules and charts.

CHART I

INTERNATIONAL TIME TABLE: TIME DIFFERENCE IN PHONE DIALING**				
TIME DIFFERENCE FROM U.S. MAINLAND TO:	**U.S. TIME ZONES IN *STANDARD TIME***			
	EASTERN (New York)	**CENTRAL (Chicago)**	**MOUNTAIN (Denver)**	**PACIFIC (Los Angeles)**
PARIS (France)	6	7	8	9
MUNICH (Germany)	6	7	8	9
ATHENS (Greece)	7	8	9	10
PEKING (China)	13	14	15	16
TEL AVIV (Israel)	7	8	9	10
TOKYO (Japan)	14	15	16	17
AUCKLAND (New Zealand)	18	19	20	21
DURBAN (South Africa)	7	8	9	10
STOCKHOLM (Sweden)	6	7	8	9
LONDON (England)	5	6	7	8
CARACAS (Venezuela)	1	2	3	4

**To compute time change, add the number of hours shown to your watch.

Directions: Answer each question by completing the sentence, choosing true or false, or selecting the best multiple-choice response.

1. There are four United States time zones listed on Chart I: Eastern, _____, _____, and Pacific.

2. Chart I shows the _____ difference between United States cities and foreign cities.

3. The time difference between Chicago and Tokyo is nineteen hours.   True   False

4. When it is 3 o'clock p.m. in New York, it is 8 o'clock p.m. in London.   True   False

5. When it is 12 o'clock noon Tuesday in Denver, it is 8 o'clock _____ in Auckland, New Zealand.
   a) Tuesday morning
   b) Tuesday evening
   c) Wednesday evening
   d) Monday morning
   e) Wednesday morning

6. When driving from one United States time zone to another, you need to change your watch by _____.
   a) three hours
   b) two hours
   c) one hour
   d) four hours
   e) five hours

SCHEDULE I

WEIGHT IN POUNDS (lbs.)	PARCEL POST RATE SCHEDULE (for packages over 1 pound)*							
	ZONES BASED ON MILES PARCEL IS SHIPPED							
	MILES	up to 150	151-300	301-600	601-1000	1001-1400	1401-1800	1801+
	ZONE: Local	1 & 2	3	4	5	6	7	8
1— 1lb.15oz.	$1.52	$1.55	$1.61	$1.70	$1.83	$1.99	$2.15	$2.48
2— 2lb.15oz.	1.58	1.63	1.73	1.86	2.06	2.30	2.55	3.05
3— 3lb.15oz.	1.65	1.71	1.84	2.02	2.29	2.61	2.94	3.60
4— 4lb.15oz.	1.71	1.79	1.96	2.18	2.52	2.92	3.32	4.07
5— 5lb.15oz.	1.78	1.87	2.07	2.33	2.74	3.14	3.64	4.54
6— 6lb.15oz.	1.84	1.95	2.18	2.49	2.89	3.38	3.95	5.02
7— 7lb.15oz.	1.91	2.03	2.30	2.64	3.06	3.63	4.27	5.55
8— 8lb.15oz.	1.97	2.11	2.41	2.75	3.25	3.93	4.63	6.08
9— 9lb.15oz.	2.04	2.19	2.52	2.87	3.46	4.22	5.00	6.62
10—10lb.15oz.	2.10	2.28	2.60	3.00	3.68	4.51	5.38	7.15
11—11lb.15oz.	2.17	2.36	2.66	3.10	3.89	4.80	5.75	7.69
12—12lb.15oz.	2.21	2.41	2.72	3.19	4.02	4.96	5.95	7.97
13—13lb.15oz.	2.26	2.46	2.78	3.28	4.13	5.12	6.14	8.24
14—14lb.15oz.	2.31	2.51	2.83	3.36	4.25	5.26	6.32	8.48
15—15lb.15oz.	2.36	2.56	2.89	3.44	4.35	5.40	6.49	8.72
16—16lb.15oz.	2.40	2.59	2.94	3.51	4.45	5.53	6.65	8.94
17—17lb.15oz.	2.44	2.64	2.99	3.59	4.55	5.65	6.80	9.15
18—18lb.15oz.	2.48	2.68	3.04	3.66	4.64	5.77	6.94	9.35
19—19lb.15oz.	2.52	2.72	3.10	3.73	4.73	5.89	7.09	9.55

*Consult postmaster for parcels over 20 pounds.
Source: U.S. Postal Service, 1981.

Directions: Answer each question by completing the sentence, choosing true or false, or selecting the best multiple-choice response.

7. Parcel post rates are determined by the number of _____ the parcel is shipped and the _____ of the parcel.

8. Schedule I gives the cost of mailing a package that weighs at least _____ pound(s) but not more
(number)
than _____ pounds.
(number)

9. According to Schedule I, mailing a seven-pound package   True   False
to a city 850 miles away costs $3.06.

10. The postage for mailing a package to a city 350 miles   True   False
away is listed under Zone 5.

11. The cost per pound to mail a ten-pound package 510 miles is _____.

    a) 85¢
    b) 30¢
    c) 35¢
    d) 17¢
    e) 28¢

12. The cost of mailing two five-pound packages 525 miles is _____ more than mailing one ten-pound package the same distance.

    a) $1.64
    b) $5.00
    c) $3.32
    d) $1.66
    e) $2.50

SCHEDULES AND CHARTS SKILLS INVENTORY CHART

Use this inventory to see what you already know about schedules and charts and what you need to work on. A passing score is 9 correct answers. Even if you have a passing score, circle the number of any problem that you miss, correct it, and turn to the practice page indicated.

Circle the Questions Missed	Type	Type of Question	Practice Page
1, 2	Chart	Scanning	72
7, 8	Schedule		72
3	Chart	Reading	72
9, 10	Schedule		72
4, 5, 6	Chart	Comprehension	72
11, 12	Schedule		72

WHAT ARE SCHEDULES AND CHARTS?

Schedules and charts show specific values by listing numbers and words in columns and rows.

COMPARING SCHEDULES AND CHARTS

Look at the examples of a schedule and a chart below. Schedules and charts may look very much alike. However, they may have different uses.

Example: Bus Schedule

Bus Schedule
#22 LCC EXPRESS

LEAVE 10th & Willamette	19th & Pearl	30th & Hilyard	ARRIVE Lane Commnty College
7:05	7:09	7:13	7:21
7:35	7:39	7:43	7:51
8:05	8:09	8:13	8:21
8:35	8:39	8:43	8:51
9:05	9:09	9:13	9:21
9:35	9:39	9:43	9:51
10:05	10:09	10:13	10:21
10:35	10:39	10:43	10:51
11:05	11:09	11:13	11:21
11:35	11:39	11:43	11:51
12:05	12:09	12:13	12:21
12:35	12:39	12:43	12:51
1:05	1:09	1:13	1:21
1:35	1:39	1:43	1:51
2:05	2:09	2:13	2:21
2:35	2:39	2:43	2:51
3:05	3:09	3:13	3:21
3:35	3:39	3:43	3:51
4:05	4:09	4:13	4:21
4:35	4:39	4:43	4:51

Source: Lane Transit District, 1982.

Schedules often list times of events. Therefore, a useful definition of a schedule is:

> a list of facts and relations primarily dealing with <u>times</u> of events.

Examples of other schedules:

train
television
sports events
school registration

Example: Nutrition Chart

Nutrition Chart of Food Groups for 3 ounce servings

FOOD	CALORIES	PROTEIN (g.)
CHICKEN	116	20
LAMB	348	17
BEEF	165	25
HAM	245	18
PORK	339	21

Source: U.S. Department of Agriculture Bulletin #72, 1978.

Charts are used to compare values of one item with another. Therefore, a useful definition of a chart is:

> a list of information on which items are <u>compared</u>.

Examples of other charts:

calorie
weight
mileage
medical costs

PARTS OF SCHEDULES AND CHARTS

1)   Schedules and Charts Have Titles

The title is a short description of the topic or main idea.

Example: Bus Schedule.

> **Bus Schedule** ♿
> **#22 LCC EXPRESS**

The *title* gives the bus route number (22) and the bus route name (LCC Express).

Example: Nutrition Chart.

> **Nutrition Chart of Food Groups**
> **for 3 ounce servings**

The *title* gives the topic: nutrition of food groups.

2)   Schedules and Charts Often Contain Tables

A table is a list of words and numbers written in rows and columns. Columns are read up and down. Rows are read across. (You will learn how to read tables on pages 74 and 75.)

LEAVE 10th & Willamette	19th & Pearl	30th & Hilyard	ARRIVE Lane Commnty College
7:05	7:09	7:13	7:21
7:35	7:39	7:43	7:51
8:05	8:09	8:13	8:21
8:35	8:39	8:43	8:51
9:05	9:09	9:13	9:21
9:35	9:39	9:43	9:51
10:05	10:09	10:13	10:21
10:35	10:39	10:43	10:51
11:05	11:09	11:13	11:21
11:35	11:39	11:43	11:51
12:05	12:09	12:13	12:21
12:35	12:39	12:43	12:51
1:05	1:09	1:13	1:21
1:35	1:39	1:43	1:51
2:05	2:09	2:13	2:21
2:35	2:39	2:43	2:51
3:05	3:09	3:13	3:21
3:35	3:39	3:43	3:51
4:05	4:09	4:13	4:21
4:35	4:39	4:43	4:51

Source: Lane Transit District, 1982.

FOOD	CALORIES	PROTEIN (g)
CHICKEN	116	20
LAMB	348	17
BEEF	165	25
HAM	245	18
PORK	339	21

Source: *U.S. Department of Agriculture Bulletin #72, 1978.*

This *table* compares nutritional values of five food groups.

This *table* gives the times when the bus leaves and arrives at specific locations.

3)   Schedules and Charts Often Have Symbols

Symbols provide additional information. A key is often used to give the meaning or value of a symbol.

♿   This symbol means that the bus has a lift to pick up riders in wheelchairs.

(g)   This symbol stands for the word "grams," the weight unit that is used to measure protein.

TYPES OF QUESTIONS

In this section of the workbook, you will be asked three types of questions. These questions will help you to find information contained on charts and schedules.

1) Scanning the Chart (Schedule) Questions

"Scanning the Chart (Schedule) Questions" require looking at the main topics and information on a chart or schedule. These questions are answered by filling in words to complete the sentence. In scanning a chart (schedule):

- Pay attention to the titles, names of columns and rows, and identified units of measurement.
- Check the source of the chart's (schedule's) information.

Example: Chart A gives the recommended daily dietary allowance of Vitamin _____ .

Answer: C. Scan the titles of the columns for this information.

2) Reading the Chart (Schedule) Questions

"Reading the Chart (Schedule) Questions" involve locating specific information on the chart or schedule. These questions will require you to answer either "true" or "false." When reading a chart (schedule):

- Read horizontally (from left to right) and vertically (from top to bottom) to find information.

CHART A

	Age (years)	Weight (lb)	Height (in)	Protein (g)	Vitamin C (mg)
RECOMMENDED DAILY DIETARY ALLOWANCES (revised 1980)					
Infants	0.0–0.5	13	24	kg × 2.2	35
	0.5–1.0	20	28	kg × 2.0	35
Children	1–3	29	35	23	45
	4–6	44	44	30	45
	7–10	62	52	34	45
Males	11–14	99	62	45	50
	15–18	145	69	56	60
	19–22	154	70	56	60
	23–50	154	70	56	60
	51+	154	70	56	60
Females	11–14	101	62	46	50
	15–18	120	64	46	60
	19–22	120	64	44	60
	23–50	120	64	44	60
	51+	120	64	44	60
Pregnant				+30	+20
Lactating				+20	+40

Reproduced from *Recommended Dietary Allowance, 1980 Ninth Edition* by permission of the National Academy Press, Washington, D.C.

Example: It is recommended that a boy, age 12, consume 56 grams of protein daily.   True   False

Answer: **False.** Starting from the left side of the chart and at the row title "Males 11–14," read across to 45 in the protein column. (12 year olds are in the 11–14 age category.)

3) Comprehension Questions

Comprehension questions require you to compare values, make inferences, and draw conclusions. You should choose the best of five possible answers.

Example: Women, 15–18 years old, require _____ mg of Vitamin C than women 11–14 years old.

Answer: a) 10 more. Subtract the amount of Vitamin C for women 11–14 (50 mg) from the number for women 15–18 (60 mg).

a) 10 more
b) 10 less
c) 50 more
d) 50 less
e) 60 more

USE CARE IN READING QUESTIONS

Careful reading may help you to answer "trickier" questions. The hints below will alert you to possible problem areas.

1) Information Is True But Not Contained On The Chart

Sample Question:

According to Chart B, milk is more nutritional than candy.   True   False

Answer: False. While this fact is true, the chart does not compare the nutritional content of foods. Chart B only recommends the number of servings for certain food groups.

2) Words That Are Incorrectly Used As Possible Answers

Sample Question:

Chart B tells _____.
- a) the percentage of certain food groups recommended for different types of people.
- b) the number of servings per food group recommended for different types of people.
- c) the percentage of servings per food group recommended for different types of people.

Answer: b) The chart is about numbers of servings. Nothing is said about percentages, so choices "a" and "c" can be eliminated.

3) Symbols, Equivalents, And Abbreviations Are Often Used To Get The Correct Answer

Sample Question:

According to Chart B, an adult requires at least _____ ounces of meat per day.

Answer: four. The chart contains two important facts. One fact is that an adult requires at least two servings of meat or a protein equivalent. The other is that a serving consists of two ounces. Therefore, two servings of two ounces gives a total of four ounces.

CHART B

Guide to Good Eating...

A Recommended Daily Pattern

The recommended daily pattern provides the foundation for a nutritious, healthful diet.

The recommended servings from the Four Food Groups for adults supply about 1200 Calories. The chart below gives recommendations for the number and size of servings for several categories of people.

Food Group	Recommended Number of Servings				
	Child	Teenager	Adult	Pregnant Woman	Lactating Woman
Milk 1 cup milk, yogurt, OR **Calcium Equivalent:** 1½ slices (1½ oz) cheddar cheese* 1 cup pudding 1¾ cups ice cream 2 cups cottage cheese*	3	4	2	4	4
Meat 2 ounces cooked, lean meat, fish, poultry, OR **Protein Equivalent:** 2 eggs 2 slices (2 oz) cheddar cheese* ½ cup cottage cheese* 1 cup dried beans, peas 4 tbsp peanut butter	2	2	2	3	2
Fruit-Vegetable ½ cup cooked or juice 1 cup raw Portion commonly served such as a medium-size apple or banana	4	4	4	4	4
Grain, whole grain, fortified, enriched 1 slice bread 1 cup ready-to-eat cereal ½ cup cooked cereal, pasta, grits	4	4	4	4	4

*Count cheese as serving of milk OR meat, not both simultaneously.

"Others" complement but do not replace foods from the Four Food Groups. Amounts should be determined by individual caloric needs.

B164 5 1980. Copyright © 1977, 4th Edition, National Dairy Council, Rosemont, IL 60018 All rights reserved.

SCHEDULES AND CHARTS

FINDING INFORMATION ON A TABLE

To answer questions about schedules and charts, you will need to know how to read the columns and rows on the table. Look at the chart below.

Nutrition Chart of Food Groups for 3 ounce servings		
FOOD	CALORIES	PROTEIN (g)
CHICKEN	116	20
LAMB	348	17
BEEF	165	25
HAM	245	18
PORK	339	21
Source: *U.S. Department of Agriculture Bulletin #72, 1978.*		

Follow the steps below to answer questions about charts and schedules.

Sample Question:
How many calories are contained in a three-ounce serving of ham?

Step 1. The table lists both the calories and protein found in the five types of foods. The chart's title states that these figures are given for three-ounce servings.

FOOD	CALORIES	PROTEIN (g)
CHICKEN	116	20
LAMB	348	17
BEEF	165	25
HAM	245	18
PORK	339	21

Step 2. Find the row labeled "Ham." Place a piece of paper under that row extending across the chart or use your finger as shown.

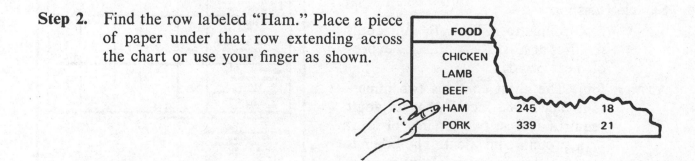

Step 3. In the column headings, locate the title "Calories." Move your eye (or finger) down that column until it meets the row labeled "Ham."

FOOD	CALORIES	PROTEIN (g)
	116	
	348	
	165	
	245	
	339	

Answer: There are 245 calories in a three-ounce serving of ham.

SAMPLE PRACTICE QUESTIONS

Now, answer the questions below. Check your answers to see that you are reading the chart correctly. Remember to carefully follow the steps shown above.

Question 1. There are seventeen grams of protein in a three-ounce serving of _____.

Question 2. A six-ounce serving of pork would contain 339 calories. True   False

Answers: **1. lamb.** In this case, you locate the information within the table and find your answer as the row title. Find seventeen in the protein column and then look across to the row labeled "Lamb."

2. False. You must multiply the given number of calories in a three-ounce serving (339) by two to get the total number of calories in a six-ounce serving. (339 × 2 = 678)

PRACTICE CHART I

Chart I is a table that contains information about the cost of dental services in various cities of the United States. This chart summarizes a survey of twelve cities.

CHART I

AVERAGE COST OF DENTAL SERVICES IN 12 UNITED STATES CITIES

Procedure	Highest Cost	Lowest Cost	12-City Average
Cleaning	$22 (San Francisco)	$12 (Cincinnati)	$18
Amalgam filling	$28 (Houston)	$19 (Cincinnati)	$23
Extraction	$22 (San Francisco)	$15 (Cincinnati)	$19
Porcelain-and-gold crown	$255 (Philadelphia)	$206 (Cincinnati)	$236
Upper denture	$347 (Houston)	$254 (Cincinnati)	$301
Bridge (Per tooth)	$247 (Philadelphia)	$197 (Cincinnati)	$225

Source: *Chicago Tribune,* March 1981.

Scanning Chart I

Fill in each blank as indicated.

1. The dental services that are listed in Chart I are: cleaning, filling, _____, crowns, _____ _____ and _____.

2. Chart I compares the highest and the lowest costs of _____ in 12 United States cities.

3. The column to the far right gives dental fees as a ____ _____ average.

Reading Chart I

Decide whether each sentence is true or false and circle your answer.

4. Of the cities listed, the city with the highest cost for cleaning is Cincinnati.   True   False

5. The twelve-city average cost for a porcelain-and-gold crown is $236.   True   False

6. The six dental procedures listed cost less in Cincinnati than in any of the other eleven areas.   True   False

Comprehension Questions

Circle the letter of the answer that best completes each sentence.

7. _____ are the two most expensive dental procedures.

   a) Dentures and crowns
   b) Bridges and cleaning
   c) Fillings and crowns
   d) Extractions and fillings
   e) Cleaning and extractions

8. The cost of a bridge (per tooth) is _____ dollars higher in Philadelphia than in Cincinnati.

   a) 7
   b) 22
   c) 28
   d) 50
   e) 95

9. The smallest dollar difference between the highest and lowest cost of any one dental procedure is for _____.

   a) fillings
   b) cleaning
   c) extractions
   d) crowns
   e) bridges

10. A conclusion that can be drawn from Chart I is:
    a) Dental procedures are less expensive in West Coast cities than in East Coast cities.
    b) The highest dental fees vary from city to city, but the lowest fees are found in Cincinnati.
    c) Dental fees have risen less during the last ten years in Cincinnati than in the other cities listed.
    d) Fees for crowns and bridges are the most expensive dental services.
    e) Cities with smaller populations tend to have lower costs for dental services.

PRACTICE SCHEDULE I

Schedule I lists registration information for classes at a community college. On the basis of the first letter of his or her last name, a student reads down the column to find the appropriate row and then reads across the row to find the day, date, and time to sign up for classes.

SCHEDULE I

TO NEW AND RETURNING STUDENTS
REGISTRATION AND INFORMATION

Registration is Monday, Jan. 26 through Saturday, Jan. 31

PLEASE NOTE FOLLOWING SCHEDULE FOR OUR
STREAMLINED REGISTRATION PROCEDURES

If Your Last Name Begins With The Letter:	Please Register On:	During The Hours of:
S·T	Monday, January 26	9:00 a.m.·1:00 p.m.
C·D·E	Monday, January 26	1:00 p.m.·5:00 p.m.
O·P·Q·R	Tuesday, January 27	9:00 a.m.·1:00 p.m.
F·G·H	Tuesday, January 27	1:00 p.m.·5:00 p.m.
M·N	Wednesday, January 28	9:00 a.m.·1:00 p.m.
I·J·K·L	Wednesday, January 28	1:00 p.m.·5:00 p.m.
U·V·W·X·Y·Z	Thursday, January 29	9:00 a.m.·1:00 p.m.
A·B	Thursday, January 29	1:00 p.m.·5:00 p.m.

Open registration regardless of first letter of last name available daily between 5 and 6 p.m. and all day Friday, Jan. 30 and Saturday, Jan. 31 from 10 a.m. to 2 p.m.

JOHNSON COMMUNITY COLLEGE
927 WEBSTER ST.

FOR FURTHER INFORMATION PHONE
787-1984

Scanning Schedule I

Fill in each blank as indicated.

1. This schedule is useful for students enrolling in classes at
   _____.

2. Student registrations begin on _____
   and end on _____.
   (day—date)   (day—date)

3. To find out more information about registration, a student can call the phone number _____.

Reading Schedule I

Decide whether each question is true or false and circle your answer.

4. A new student, Henry Foster, can register on Tuesday, January 27, between 1:00 p.m. and 5:00 p.m.   True   False

5. Mercedes Rodriquez can register on Thursday before 5:00 p.m.   True   False

6. If Nate Fennerty does not register by Tuesday at 5:00 p.m., he will not be able to register during the semester.   True   False

Comprehension Questions

Circle the letter of the answer that best completes each sentence.

7. Until 5:00 p.m., the total number of registration hours on Monday is _____.
   a) 4
   b) 8
   c) 9
   d) 10
   e) none of the above

8. If a student misses his or her assigned registration time, the student can register _____.
   a) between 5-6:30 p.m. daily
   b) on Saturday, January 31
   c) on Friday, January 30
   d) b and c
   e) a, b, and c

9. Sue Smith will have _____ hours in which to enroll on January 26.
   a) 1
   b) 3
   c) 5
   d) 8
   e) 9

10. If 160 students enroll by 1:00 p.m. on Monday, the average number of registrations per hour would be _____.
    a) 70
    b) 40
    c) 160
    d) 10
    e) 80

PRACTICE SCHEDULE II

Schedule II lists times that AMTRAK (the National Railroad Passenger Corporation) trains arrive at and leave cities between Sacramento and Los Angeles, California.

For ease of reading, origin and destination stations are indicated with a "Dp" for "departs" and an "Ar" for "arrives." In addition, any station that has a scheduled "dwell time" also has an "Ar" and a "Dp" indicator to tell the reader how long the train is scheduled to remain in that station. At stations without an "Ar" or "Dp" indicator, the train stops only long enough to permit passengers to board safely.

Notice that small picture symbols tell the services available on this train. For example, ▭ means that this train has sandwich, snack, and beverage services.

SCHEDULE II

Western Schedules

Sacramento-Oakland-Los Angeles

READ DOWN			READ UP
15	Train Number		18
Daily	Frequency of Operation		Daily
🔲 🛏 ⊠ ⬭			🔲 🛏 ⊠ ⬭
7 55 P	Dp	Sacramento, CA	Ar	9 30 A
8 13 P		Davis, CA		8 48 A
8 41 P		Suisan-Fairfield, CA		8 21 A
9 03 P		Martinez, CA		7 59 A
9 32 P		Richmond, CA		7 28 A
9 50 P	Ar	Oakland, CA	Dp	7 15 A
10 10 P	Dp	Oakland, CA	Ar	7 00 A
11 25 P	Ar	San Jose, CA	Dp	5 45 A
11 30 P	Dp		Ar	5 41 A
12 55 A		Salinas, CA		4 12 A
3 51 A	Ar	San Luis Obispo, CA	Dp	1 28 A
3 59 A	Dp	(Hearst Castle)	Ar	1 20 A
6 17 A		Santa Barbara, CA		10 45 P
7 06 A		Oxnard, CA		9 53 P
8 18 A		Glendale, CA		8 43 P
9 00 A	Ar	Los Angeles, CA	Dp	8 25 P

Registered U.S. Patent and Trademark Office.

Reference Marks

🔲 All reserved train.
🛏 Sleeping car service.
⬩ Club car service.

⊠ Tray meal and beverage service.
▭ Sandwich, snack and beverage service.
⬭ Checked baggage handled.

A   a.m.
P   p.m.
Ar Arrive
Dp Depart

Amtrak ➤ *

Scanning Schedule II

Fill in each blank as indicated.

1. The train stops at four cities between Sacramento and
   Oakland: Davis, _____,
   _____, and _____.

2. The number of the train going from Sacramento to Los Angeles is _____, and the number of the train going from Los Angeles to Sacramento is _____.

3. The symbol 🖐 means that the train has _____ _____.

Reading Schedule II

Decide whether each sentence is true or false and circle your answer.

4. Traveling from Sacramento, train #15 arrives in Los Angeles at 9:00 a.m. the next day.   True   False

5. A train to Los Angeles leaves San Jose at 7:00 a.m.   True   False

6. Train #18 has sleeping car service.   True   False

Comprehension Questions

Circle the letter of the answer that best completes each sentence.

7. The total time it takes to travel from Sacramento to Oakland is slightly less than _____.

    a) 4 hours
    b) 9 hours
    c) 7 hours
    d) 2 hours
    e) 1 hour

8. If you leave Los Angeles on Friday evening at 8:25 p.m., you will arrive in Sacramento at 9:30 _____.

    a) Saturday evening
    b) Sunday morning
    c) Friday evening
    d) Saturday morning
    e) Sunday evening

9. Train #18 stops at Oakland for _____ minutes before it leaves for Sacramento.

    a) 30
    b) 15
    c) 20
    d) 10
    e) 12

10. It takes a little over _____ for train #15 to travel from Sacramento to Los Angeles.

    a) 25 hours
    b) 13 hours
    c) 5 hours
    d) 8 hours
    e) 2 hours

APPLYING YOUR SKILLS: SCHEDULES AND CHARTS

A single person must file a tax return, known as Form 1040, in each year that his/her income is at least $3,300 for the year. Likewise, a married person filing a return with his/her spouse must file if their income is at least $5,400.

Many people choose to have their taxes prepared by a professional. Chart I compares the times for preparation, quoted fees, and actual fees of various tax preparers. In addition, Chart I includes the total federal and state tax bill, which is the amount of tax each individual taxpayer, based on his income, pays to the state and federal governments.

WHO WILL DO YOUR TAX RETURN?

From January to April, newspapers are filled with advertisements by income tax preparers. Both prices and the amount of money a tax preparer might save you vary widely...

CHART I

Comparing tax preparers

Tax preparation firm or agency	Quote fee	Actual fee	Total federal and state tax bill	Time from appointment to receiving return	Was it longer than they estimated?
Butler & Associates, Inc.	$40-45	$130	$6,822	Same day	Same day
H&R Block, 7 W. Madison St.	38-40	96	6,160	19 days	9 days late
Lapat & Sokolow, Ltd., attorneys	200	200	6,462	16 days	6 days late
Dutton & Co., Certified Public Accountant	80	80*	6,207**	33+ days	17+ days late
Atlas Insurance Agency	25-30	75	6,840	4 days	On time
Beneficial Income Tax Service	45-50	68	6,301	17 days	9 days late
Frank T. Kapple, Certified Public Accountant	125	120	6,219	14 days	2 days early
Ernst & Whinney, accountants	300-400	375*	6,290	9 days	12 days early
Internal Revenue Service	0	0	7,155	Same day	Same day
Deloitte Haskins & Sells, accountants	250-400	350*	6,206	23 days	6 days late
Jay A. Slutzky, attorney at law	25-50	100	6,673	7 days	5 days early
H&R Block, 1352 E. 53d St.	30-50	79	6,291	20 days	11 days late

*Estimate
**Preliminary figure

Source: Copyrighted, 1981, *Chicago Tribune.* Used with permission.

Directions: Answer each question by completing the sentence, choosing true or false, or by selecting the best multiple-choice response.

1. Chart I compares quoted fees with _____ fees charged by income tax preparers.

2. Chart I contains a list of 12 firms whose business is

   _____

3. From Chart I, it can be concluded that many tax preparers are late in completing their client's tax returns.

True   False

4. The Internal Revenue Service does not charge for tax preparation assistance.

True   False

5. For most firms listed, actual fees are less than quoted fees.

True   False

6. The two firms that provide same day service are _____.

a) Butler and Atlas
b) IRS and Slutzky
c) Kapple and IRS
d) Butler and IRS
e) Butler and Dutton

7. According to Chart I, the difference between the quoted fee and the actual fee charged by H&R Block, 7 W. Madison Avenue, ranges from _____.

a) $29 to $49
b) $37 to $67
c) $56 to $58
d) $38 to $68
e) none of the above

8. If you had an appointment with the Atlas Insurance Agency on Friday, March 12th, you could expect your tax return by _____.
(Note: Days given on chart are working days.)

a) Friday, March 12
b) Tuesday, March 16
c) Thursday, March 18
d) Friday, March 19
e) Tuesday, March 23

9. The diagram that represents Chart I most accurately is:

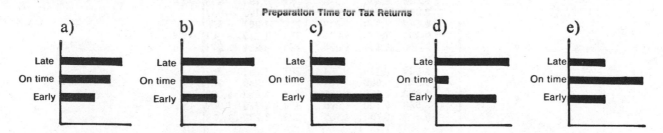

10. The statement that best describes Chart I is:
a) All tax preparation firms charge more than their quoted fees.
b) IRS does not charge for tax preparation assistance, but it calculates the highest tax bill.
c) Tax preparation firms vary widely in fees charged, total tax bill, and time to prepare forms.
d) The tax preparation firms that charge the least for their service save you the most money.
e) Attorneys and accountants provide the best tax service.

A tax table is a schedule that lists the tax to be paid by a wage earner. This tax is based on total income earned and filing status. Income tax rates are based on the principle of "graduated income tax." This principle means that the amount of tax to be paid increases as income increases.

WILL YOU PAY OR GET SOMETHING IN RETURN?
From January to April each year, people hope for the return of money from the Internal Revenue Service. On the other hand, many fear that the taxes they've already paid may not have been enough. As a result, they may have to pay more income tax. The IRS reports that . . . .

SCHEDULE I

1981 Tax Table (Continued)

If line 12 (taxable income) is —		And you are—			
At least	But less than	Single	Married filing jointly	Married filing separately	Head of a household
			Your tax is—		
8,000					
8,000	8,050	969	698	1,124	893
8,050	8,100	979	707	1,136	902
8,100	8,150	988	715	1,148	911
8,150	8,200	998	724	1,160	920
8,200	8,250	1,007	733	1,172	929
8,250	8,300	1,016	742	1,184	938
8,300	8,350	1,026	751	1,195	947
8,350	8,400	1,035	760	1,207	955
8,400	8,450	1,045	769	1,219	964
8,450	8,500	1,054	778	1,231	973
8,500	8,550	1,064	787	1,243	982
8,550	8,600	1,074	795	1,255	991
8,600	8,650	1,085	804	1,266	1,000
8,650	8,700	1,095	813	1,278	1,009
8,700	8,750	1,105	822	1,290	1,019
8,750	8,800	1,116	831	1,302	1,029
8,800	8,850	1,126	840	1,314	1,040
8,850	8,900	1,136	849	1,326	1,051
8,900	8,950	1,147	858	1,338	1,062
8,950	9,000	1,157	867	1,349	1,073

Directions: Answer each question by completing the sentence, choosing true or false, or selecting the best multiple-choice response.

1. The portion of the tax table shown applies to an income equal to or greater than $_____ but less than $_____.

2. The income tax paid is based on one of the four types of marital status, namely:

1) _____

2) _____

3) _____

4) _____

3. On the IRS form accompanying the tax table, line 12 represents taxable income.

True   False

4. The tax paid by a single woman on an income of $8,225 per year is $1,016.

True   False

5. A man who earns $8,950 a year and pays more than $1,200 in taxes, falls into the category of "Married-Filing Separately."

True   False

6. A single person earning $8,000 a year pays _____ less taxes than a single person earning $8,955.

a) $195
b) $150
c) $178
d) $159
e) none of the above

7. With an income level of $8,558, a single person pays _____ than a married person filing jointly with a spouse.

a) $279 more
b) $289 more
c) $279 less
d) $289 less
e) none of the above

8. Members of the filing category who pay the least tax at every income level shown on Schedule I are _____.

a) single
b) married, filing jointly
c) married, filing separately
d) heads of households
e) public aid recipients

9. All of the following statements can be concluded from the 1981 Tax Table, Schedule I, EXCEPT:

a) As income increases, total taxes paid in every filing category increase.

b) Taxes paid each year are in part dependent on marital status.

c) Married individuals, filing jointly, pay a lower amount of taxes than people in other categories.

d) As income decreases, taxes also decrease.

e) Reducing the tax rate would benefit heads of households more than any other group.

FINAL SCHEDULES AND CHARTS SKILLS INVENTORY

CHART A

UNITED STATES MILEAGE CHART

For Selected Cities

	Atlanta, Ga.	Chicago, Ill.	Dallas, Tex.	Denver, Colo.	Detroit, Mich.	Kansas City, Mo.	Los Angeles, Ca.	Louisville, Ky.	Memphis, Tenn.	Milwaukee, Wis.	Minneapolis, Minn.
Atlanta, Ga.		674	795	1398	699	798	2182	382	371	761	1068
Chicago, Ill.	674		917	996	266	499	2054	292	530	87	405
Dallas, Tex.	795	917		781	1143	489	1387	819	452	991	936
Denver, Colo.	1398	996	781		1253	600	1059	1120	1040	1029	841
Detroit, Mich.	699	266	1143	1253		743	2311	360	713	353	671
Kansas City, Mo.	798	499	489	600	743		1589	520	451	537	447
Los Angeles, Ca.	2182	2054	1387	1059	2311	1589		2108	1817	2087	1889
Louisville, Ky.	382	292	819	1120	360	520	2108		367	379	697
Memphis, Tenn.	371	530	452	1040	713	451	1817	367		612	826
Milwaukee, Wis.	761	87	991	1029	353	537	2087	379	612		332
Minneapolis, Minn.	1068	405	936	841	671	447	1889	697	825	332	

Directions: Answer each question by completing the sentence, choosing true or false, or selecting the best multiple-choice response.

1. The numbers on the chart represent _____.

2. Chart A is used for determining _____ between _____.

3. The mileage between Atlanta and Louisville is 382 miles.   True   False

4. There are 1,059 miles between Los Angeles and Dallas.   True   False

5. The mileage from Chicago to Denver is approximately _____ more than the mileage from Chicago to Memphis.

    a) 466 miles
    b) 530 miles
    c) 266 miles
    d) 704 miles
    e) 544 miles

6. The total miles driven on a trip from Dallas to Denver and then from Denver to Kansas City is _____.

    a) 1,237 miles
    b) 1,266 miles
    c) 1,381 miles
    d) 532 miles
    e) 650 miles

7. If a trip from Los Angeles to Chicago took five days of driving time, the average number of miles driven each day would be approximately _____.

    a) 2,054 miles
    b) 383 miles
    c) 125 miles
    d) 400 miles
    e) 410 miles

8. Averaging 400 miles for each tank of gas, how many times would you need to fill the tank as you drove from Kansas City to Atlanta? _____

    a) 4 times
    b) 2 times
    c) 8 times
    d) 3 times
    e) 6 times

SCHEDULE A

the leisure bus . . .	
GO TRAVEL-WAYS	
Schedule Departures	
Bustown, USA	

NORTHBOUND					SOUTHBOUND				
Portland		**Seattle**			**San Francisco**		**Los Angeles**		
A.M.	P.M.	A.M.	P.M.		A.M.	P.M.	A.M.	P.M.	
2:05	2:20	1:10	2:15		1:40	5:30	2:40	4:25	
5:25	4:50	5:30	4:50		5:20	10:15	8:20	9:30	
6:05	5:40	7:25	9:00		11:45		10:50		
7:25	7:10	9:10	11:45						
9:10	7:50	11:35							
11:35	9:00								
	10:00								
	11:45								

Travel-Ways
Bus Lines

Directions: Answer each question by completing the sentence, choosing true or false, or selecting the best multiple-choice response.

9. Schedule A tells only the times when buses _____ the Bustown, USA, bus depot.

10. The bus departure schedule tells passengers the times when buses leave northbound to the cities of _____ and _____ and southbound to the cities of _____ and _____.

11. Schedule A tells the time when buses arrive at their destinations.   True   False

12. The next bus going to Seattle after 11:45 p.m. leaves at 2:15 p.m.   True   False

13. A passenger who misses the 9:30 p.m. bus to Los Angeles must wait about _____ hours for the next one.

a) 3
b) 4
c) 5
d) 12
e) 24

14. The earliest daily bus to Los Angeles leaves the station at _____.

a) 1:20 a.m.
b) 6:30 a.m.
c) 7:10 a.m.
d) 2:40 a.m.
e) 3:30 p.m.

15. The graph that most accurately represents Schedule A is:

a)   b)   c)   d)

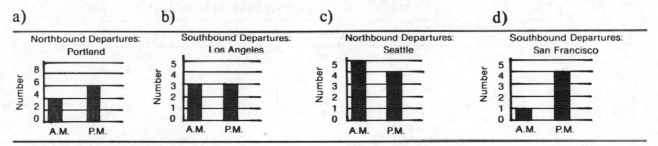

e) none of the above

16. Seattle is located north of Portland. Many of the departure times listed for Seattle are also listed for Portland. You can infer from Schedule A that:

a) One bus goes directly to Portland, while another goes directly to Seattle.

b) Some of the buses that go northbound to Seattle are the same buses that are going to Portland.

c) Due to demand, several buses leave the station at the same time.

d) There are more passengers going to Seattle than to Portland.

e) None of the above.

SCHEDULE B

Yearly Interest Rate: 13 %	MONTHLY PAYMENT SCHEDULE				
	Amount of Car Loan				
No. of Months	$1000	$2000	$3000	$4000	$5000
12	$89.32	$178.64	$267.96	$357.27	$446.59
18	$61.45	$122.90	$184.35	$245.80	$307.24
24	$47.55	$95.09	$142.63	$190.17	$237.71
30	$39.23	$78.45	$117.67	$156.89	$196.11
36	$33.70	$67.39	$101.09	$134.78	$168.47
48	$26.83	$53.66	$80.49	$107.31	$134.14

Directions: Answer each question by completing the sentence, choosing true or false, or selecting the best multiple-choice response.

17. Schedule B is a payment schedule for a _____ loan borrowed at _____ percent yearly interest rate.

18. The payment schedule applies to loans in the amount of $_____ to $_____.

19. A car loan of $3,000 borrowed for 12 months requires a monthly payment of $267.96.   True   False

20. To keep monthly payments below $60 on a $2,000 loan means borrowing the money for 48 months.   True   False

21. For a two-year loan, the difference between the monthly payments on $2,000 and $4,000 is _____.
   a) $95.09
   b) $95.08
   c) $57.55
   d) $113.77
   e) $27.46

22. For a $5,000 loan, monthly payments made over a two year period are _____ higher than monthly payments made over three years.
   a) $168.47
   b) $196.11
   c) $69.24
   d) $103.57
   e) none of the above

23. The graph that most accurately represents Schedule B is:

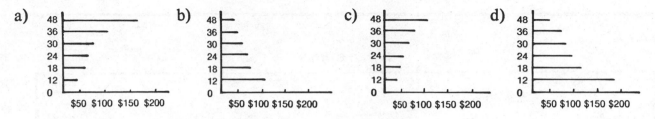

Monthly Payments on Car Loan: $2,000

e) none of the above

24. The statement that best describes the payment schedule is:
a) More loans are made for 48 months than for any other time period.
b) Monthly payments on a car are determined by the amount of the down payment and by credit status.
c) Monthly payments on a car loan are determined by interest rate, amount borrowed, and the amount of time for repayment.
d) High interest rates for car loans are due to inflation.
e) A forty-eight month loan is easier to repay than a twelve month loan because the monthly payments are less.

FINAL SCHEDULES AND CHARTS SKILLS INVENTORY CHART

Circle the number of any problem that you missed and be sure to review the appropriate page. A passing score is 19 correct answers. If you miss more than 5 questions, you should review this chapter.

Circle the Questions Missed	Type	Type of Question	Practice Page
1, 2	Chart	Scanning	72
9, 10, 11, 17, 18	Schedule		72
3, 4	Chart	Reading	72
12, 14, 19, 20	Schedule		72
5, 6, 7, 8	Chart	Comprehension	72
13, 15, 16, 21, 22, 23, 24	Schedule		72

MAPS
MAP SKILLS INVENTORY

This inventory will help you measure your skills in reading and interpreting maps. Following this skills inventory, you will find instructional material and maps for further practice.

MAP A

UNITED STATES

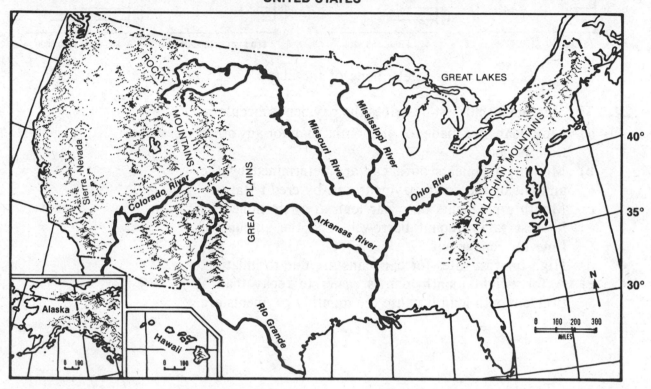

Directions: Answer each question by completing the sentence, choosing true or false, or selecting the best multiple-choice response.

1. On Map A, the symbol ![symbol] represents a _____ _____.

2. The lines that run across the United States from left to right (labeled 35°, 40°, etc.) are called _____ lines.

3. The Missouri River is in the northeastern United States.   True   False

4. The Rocky Mountains is a major mountain range in the western United States.   True   False

5. The Appalachian Mountains stretch for approximately _____ miles from the northeastern to the southeastern part of the United States.
   - a) 1,800
   - b) 2,400
   - c) 1,200
   - d) 300
   - e) 900

MAP B

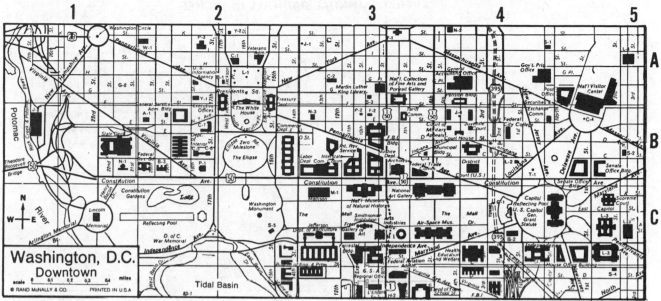

©Rand McNally & Co., R. L. 83-5-59.

Directions: Answer each question by completing the sentence, choosing true or false, or selecting the best multiple-choice response. A map-section code is given in parentheses for each place mentioned in the questions.

6. The above map shows the _____ area of Washington, D.C.

7. The Supreme Court (C-5) is located about 0.25 miles _____ of the U.S. Capitol Building. (C-4)
   (direction)

8. The Lincoln Memorial is located in map section C-2.   True   False

9. 17th Street crosses Highway 50 less than $\frac{1}{2}$ mile from the Washington Monument (C-2).   True   False

10. The distance between the U.S. Capitol Building (C-4) and the White House (B-2) is about _____.
    a) $\frac{1}{2}$ mile
    b) 1 mile
    c) $1\frac{1}{2}$ miles
    d) 2 miles
    e) $2\frac{1}{2}$ miles

11. Starting at the F.B.I. Building (B-3), walk southeast on Pennsylvania Avenue for 0.4 mile to end up near the _____.
    a) Air-Space Museum
    b) Lincoln Memorial
    c) Internal Revenue Service
    d) U.S. District Court
    e) U.S. Capitol Building

MAP C

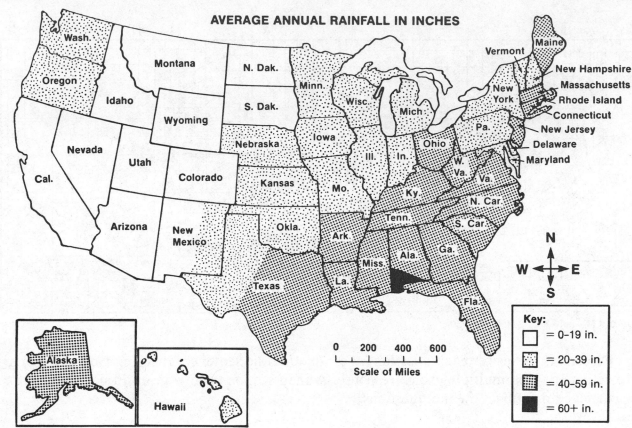

Source: National Oceanic & Atmospheric Administration, U.S. Department of Commerce, 1979.

Directions: Answer each question by completing the sentence, choosing true or false, or selecting the best multiple-choice response.

12. The scale of miles measures a total of _____ miles.

13. Map C shows the normal average _____ in _____ for the United States.

14. The symbol [▦] means that the average rainfall for a particular state is from _____ to _____ inches per year.

15. Ohio receives 40–59 inches of rain per year.   True   False

16. Only one state receives two measurably different amounts   True   False
of rainfall yearly.

17. Of the following states, the state with the least average rainfall per year is _____.

    a) Hawaii
    b) Texas
    c) Illinois
    d) Wyoming
    e) Alabama

18. Of the following states, the one shown by the map to have the driest climate is _____.

    a) Maine
    b) New York
    c) Illinois
    d) California
    e) Nebraska

MAP SKILLS INVENTORY CHART

Use this inventory to see what you already know about maps and what you need to work on. A passing score is 15 correct answers. Even if you have a passing score, circle the number of any problem that you miss, correct it, and turn to the practice page indicated.

Circle the Questions Missed	Type of Map	Type of Question	Practice Page
1, 2	Geographical	Scanning	109
6	Directional		115
12, 13, 14	Informational		121
3, 4	Geographical	Reading	109
7, 8, 9	Directional		115
15, 16	Informational		121
5	Geographical	Comprehension	109
10, 11	Directional		115
17, 18	Informational		121

WHAT ARE MAPS?

A *map* is a visual display that represents the whole earth or a particular region of it. Some maps show natural features such as land masses, mountains, rivers, and oceans. Other maps show man-made features such as boundary lines between countries or states and the location of cities and highways. Maps are also used to show special information such as weather conditions and time zones.

TYPE OF MAPS

Maps are widely used in education, business, and recreation. Because of their wide variety of uses, it is convenient to group maps into three categories for study in this book: *geographical*, *directional*, and *informational*.

Geographical Maps

A *geographical map* shows the natural features of a region of the earth. *Natural features* are the lands, rivers, lakes, oceans, and other features that are not man-made.

The illustration below is an example of a geographical map. This map of North America shows height (elevation) by darkening or shading mountainous or other elevated land areas.

Directional Maps

A *directional map* is used to show the location of cities, highways, and points of interest. Directional maps called "road maps" are commonly used by travelers to find their way across the country, across a state, or within a city. The use of these maps allows a traveler to plan a route in advance and to estimate time of travel.

The map at the right shows the main high-ways within the state of Texas.

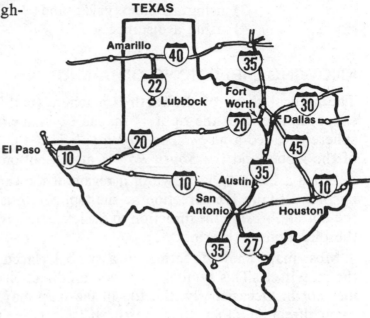

Informational Maps

An *informational map* gives specific information about a particular area, or it gives information comparing different areas.

This informational map compares the 1979 birth rates in the Canadian provinces.

SPECIAL MAP SKILLS

Before studying each type of map, we'll first learn four map skills that are useful for all maps:

1) knowing directions on a map,
2) understanding longitude and latitude lines,
3) using an index guide, and
4) using a distance scale.

KNOWING DIRECTIONS ON A MAP

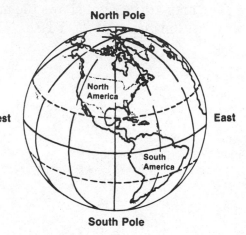

The natural shape of the earth is a sphere (ball-like shape) as shown at the right. A map in the shape of a sphere is called a _globe_. The _North Pole_ is at the top of the globe, and the _South Pole_ is at the bottom.

When a flat map is drawn of a region of the earth, _north_ is usually the direction at the top. _South_ is at the bottom, _west_ is the direction to the left, and _east_ is the direction to the right.

Most maps show direction by a symbol placed on the map itself. This symbol may show all directions or just north. Occasionally, the top of the map may be some direction other than north. In this case, the direction symbol will indicate proper directions. Shown at right are several common map direction symbols.

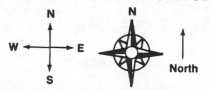

COMMON DIRECTION SYMBOLS

The four main map directions are often used in combination. For example, as the drawing at the right shows, Point A may be both north _and_ west of Point O. In this case, we say that Point A is _northwest_ of Point O. Similarly, Point B is _southeast_ of Point O. The word "north" or "south" always precedes the word "east" or "west" when directions are used in combination.

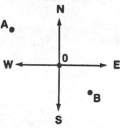

USING DIRECTIONS IN COMBINATION

The example below shows several correct uses of map directions.

EXAMPLE: At right is a map of Texas that shows the locations of several major cities.

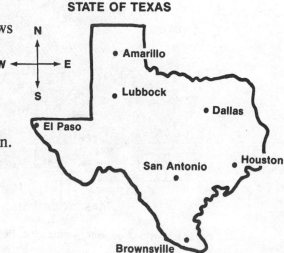

STATE OF TEXAS

a) Amarillo is _north_ of Lubbock.
b) Amarillo is _northwest_ of Dallas.
c) Amarillo is _northeast_ of El Paso.
d) Brownsville is in _southern_ Texas.
e) Brownsville is _southwest_ of Houston.
f) El Paso is in _western_ Texas and is _southwest_ of Dallas.
g) Houston is in _eastern_ Texas.
h) San Antonio is _southeast_ of Lubbock.

As a quick way for identifying parts of the country, the United States is often divided into the following six regions: Northwest, North Central, Northeast, Southwest, South Central, and Southeast.

Become familiar with these regions before answering the questions below.

Directions: To check your understanding of map directions, fill in each blank below with one of the following directions: north, south, east, west, northeast, northwest, southeast, or southwest.

1. Nevada is _____ of New Mexico.

2. Illinois is _____ of Minnesota.

3. To find Maine, look _____ of Ohio.

Answer the following questions about the six regions of the United States.

4. The largest region on the West Coast is the _____.

5. The region that contains the least number of states is the _____.

6. Florida is in the region known as the _____.

UNDERSTANDING LONGITUDE AND LATITUDE LINES

To identify the location of a point on the globe, mapmakers draw two sets of lines called longitude and latitude lines.

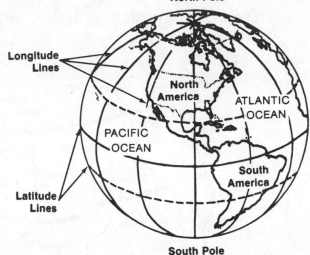

Longitude lines are drawn from the top of the globe at the North Pole to the bottom of the globe at the South Pole. These lines are often called North-South lines. Notice that longitude lines meet at each pole.

Latitude lines are drawn across the globe from left to right. These lines are often called East-West lines. Notice that latitude lines are parallel and circle the earth; they never meet at a point and they never cross.

Longitude and latitude lines are numbered in units called degrees. The symbol for degrees is "°". Latitude lines are numbered from 0° to 90°. The latitude line numbered 0° is a special line called the _equator._ The equator is halfway between the North and South Poles, and it divides the globe into two parts called _hemispheres._

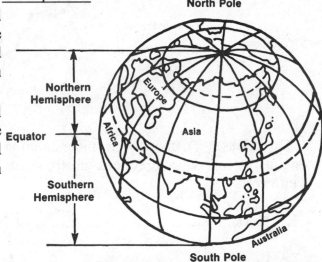

The _Northern Hemisphere_ consists of all places that lie north of the equator. Latitude lines in the northern hemisphere are numbered from 0° at the equator to 90° at the North Pole.

The _Southern Hemisphere_ consists of all places that lie south of the equator. Latitude lines in the southern hemisphere are numbered from 0° at the equator to 90° at the South Pole.

Longitude lines are numbered from 0° to 180°. The longitude line numbered 0° passes through Greenwich, England. From Greenwich, the longitude lines are numbered from 0° to 180° as you move toward the west, and from 0° to 180° as you move toward the east. The longitude line numbered 180° is exactly halfway around the earth from Greenwich, England.

The continents are also often said to be in the _Western Hemisphere_ or in the _Eastern Hemisphere._ The globe at the top of the page shows the continents in the Western Hemisphere: North and South America. The globe in the lower half of the page shows the continents of the Eastern Hemisphere: Europe, Asia, Africa, and Australia. The continent of Antarctica is located in the Southern Hemisphere.

On the next page, all the continents of the earth are shown on a single flat map. This map is called a "Mercator Projection." Notice that the longitude lines are drawn to be parallel to each other. Although this map shows the exact locations of land and water areas of the earth, it does distort their actual sizes as you look to the far north and far south. In particular, Greenland appears to be much larger on a Mercator Projection than it is on a globe.

Become familiar with this map of the earth and then answer the questions below.

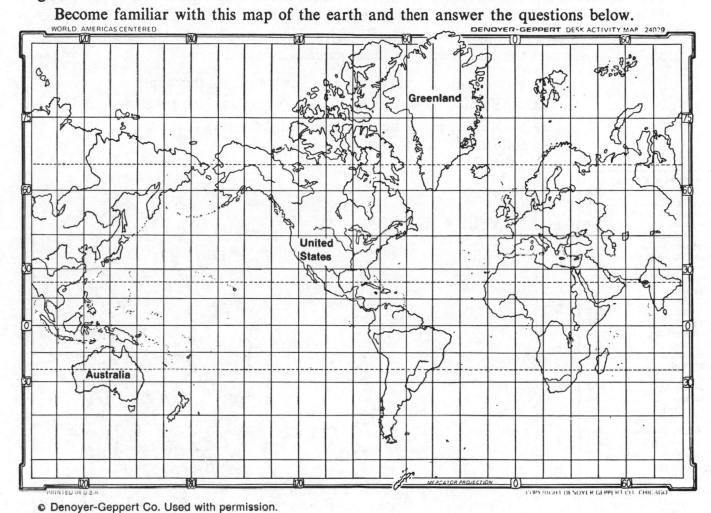

© Denoyer-Geppert Co. Used with permission.

Directions: To check your understanding of longitude and latitude lines, answer each question below.

1. The lines that run north and south are called _____ lines.

2. The lines that run east and west are called _____ lines.

3. The latitude line that is numbered 0° is called _____.

4. The latitude line that runs close to the southern border of the United States is the _____ degree line.

5. Australia lies completely _____ of the equator.
   (direction)

USING AN INDEX GUIDE

On most maps, an *index guide* is used to locate countries, cities, and other points of interest. While longitude and latitude lines identify a point's location on any globe or map, an index guide locates a point only on the particular map being used.

An index guide is a listing that identifies a small area on a map by which a specific place is located. As seen on the map below, an index guide consists of:

1) a row of numbers across the top (or bottom) of the map,
2) a column of letters along the left (or right) side of the map, and
3) a listing of places beside (or beneath) the map that are followed by a letter-number code.

Countries of Europe
Albania G-6
Austria E-6
Belgium................E-4
Bulgaria................F-7
Czechoslovakia E-6
Denmark D-5
East Germany E-5
Finland B-7
France E-3
Greece G-7
Hungary F-6
Ireland.................D-2
Italy F-5
Netherlands D-4
Norway C-5
Poland..................D-6
Portugal F-1
Rumania F-7
Spain...................F-2
Sweden.................B-6
Switzerland F-4
United Kingdom D-3
U.S.S.R. (Russia) D-7
West Germany E-5
Yugoslavia F-6

Major Cities
Amsterdam (Neth).....D-4
Athens (Gr) G-7
Barcelona (Sp) F-3
Berlin (EG) D-5
Bonn (WG) E-4
Brussels (Bel) E-4
Copenhagen (Den) D-5
Dublin (Ire)............D-3
Hamburg (WG) D-5
Helsinki (Fin) C-7
Lisbon (Por) F-1
Liverpool (UK) D-3
London (UK)............D-3
Lyon (Fr) F-4
Madrid (Sp)............F-2
Marseille (Fr) F-4
Milan (It) F-4
Munich (WG)...........E-5
Oslo (Nor)..............C-5
Paris (Fr) E-4
Rome (It) G-5
Stockholm (Swe) C-6
Vienna (Aus) E-6
Zurich (Swe) E-4

Major Rivers
Danube E-5
Ebro....................F-2
Elbe.....................D-5
Loire E-3
Po.......................F-4
Rhine...................E-4
Rhone...................F-4
Seine....................E-4
Tajo F-2

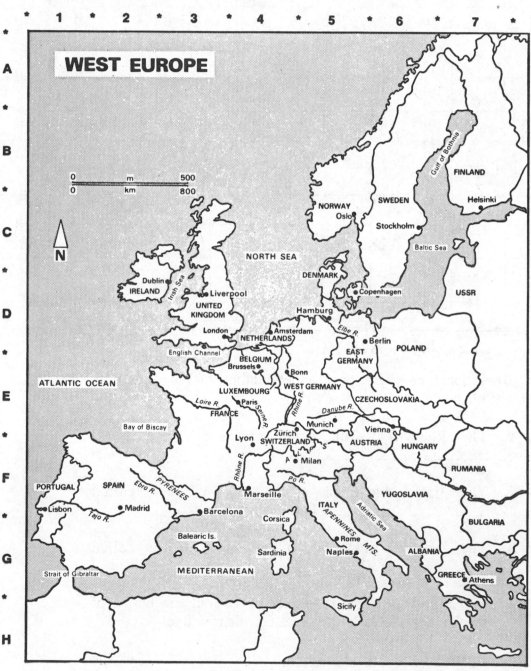

The example below shows the correct use of an index guide.

EXAMPLE: Locate the city of Helsinki.

Step 1. Find the name of the city Helsinki on the index guide.
Read the letter-number code: C-7

Step 2. Locate "C" on the left side of the map. Scan directly to the right, and stop below the number "7." The point of intersection of the two lines of sight (across from C and below 7) gives the area in which Helsinki is located.

Answer: **Helsinki is on the southern coast of Finland.**

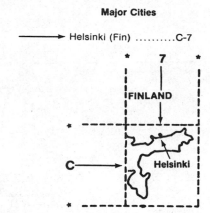

Major Cities

Helsinki (Fin) C-7

Notice that the letters and numbers are centered in the area they refer to. For instance "C" refers to the area halfway up to "B" and halfway down to "D". The boundaries are marked with a star (*).

Directions: Check your understanding of the use of an index guide. Using the map on the previous page, answer each question below by completing the sentence or by choosing true or false.

1. Write the letter-number code of each of the following cities:

   a) Oslo _____
   b) Athens _____
   c) Bonn _____

2. The Loire River is in the country of _____.

3. The large body of water in map section C-6 is the ____

   _____.

4. Near what large city does the Elbe River run into the North Sea?

   _____.

5. In what country is Lisbon?

   _____.

6. The largest country that shares a border with Poland is Russia (U.S.S.R.).   True   False

7. Stockholm is north of Oslo.   True   False

8. Marseilles is in map section F-2.   True   False

USING A DISTANCE SCALE

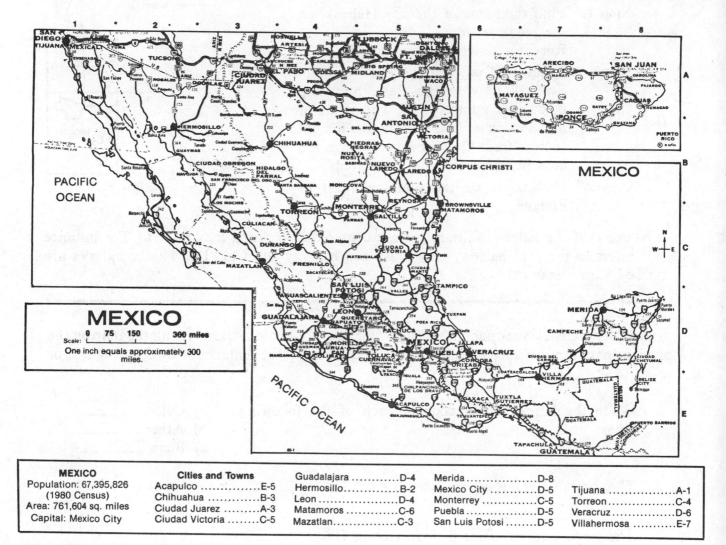

MEXICO
0   75   150   300 miles
Scale:
One inch equals approximately 300 miles.

MEXICO	Cities and Towns		
Population: 67,395,826 (1980 Census) Area: 761,604 sq. miles Capital: Mexico City	Acapulco ....E-5 Chihuahua ....B-3 Ciudad Juarez ....A-3 Ciudad Victoria C-5	Guadalajara ....D-4 Hermosillo..............B-2 Leon ....D-4 Matamoros ....C-6 Mazatlan.................C-3	Merida ....D-8 Mexico City ....D-5 Monterrey ....C-5 Puebla ....D-5 San Luis Potosi D-5	Tijuana ....A-1 Torreon.................C-4 Veracruz................D-6 Villahermosa E-7

© Rand McNally & Co., R. L. 83-5-59.

EXAMPLE: What is the direct (by air) distance between Torreon and Matamoros?

Step 1. Using the index guide, find the two cities: Torreon and Matamoros.

Step 2. Using a ruler, measure the approximate distance between them: about 1½ in.

Step 3. Multiply 1½ by 300, the number of miles per inch as given by the distance formula (found under the map title).

$$300 \times 1\tfrac{1}{2} = 300 \times \tfrac{3}{2} = 450 \text{ miles}$$

Answer: **About 450 miles**

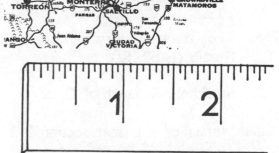

As the previous example shows, you can use a ruler to measure the *direct distance* between two points. However, the direct distance is not the *road distance.* The actual road distance between two points is always greater than the direct distance measured by ruler. In the previous example, the actual road distance between Torreon and Matamoros is about 575 miles. Actual road mileage between two cities is often written on the map itself beside the road connecting the cities.

When you don't have a ruler available, you can use a second method to measure direct distance. You can make a *mileage ruler* by drawing a distance scale on a piece of paper. The example below shows this second method.

EXAMPLE: Find the direct distance between the city of Mexico (Mexico City) and Merida.

  Step 1.   Using the index guide, find the two cities: Mexico and Merida.

  Step 2.   Use the distance scale to make a mileage ruler. Then, measure the direct mileage between the cities.

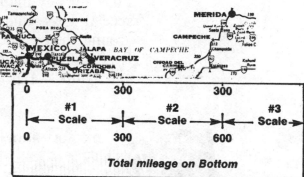

As seen to the right, the mileage ruler needed is made from 3 mileage scales placed end to end. The total mileage between the cities is just a little over 600 miles.

  Answer: **The distance is about 610 miles.**

Directions: To check your understanding of the use of a distance scale, answer each question below. Use a ruler or make a mileage ruler to measure direct distances on the map on page 104.

1.   What is the direct mileage between each pair of cities below? Round off to the nearest 50 miles.

  a) Veracruz - Chihuahua _____   c) Mazatlan - Hermosillo _____

  b) Mexico - Villahermosa _____   d) Guadalajara - Acapulco _____

2.   Approximately how close is Mexico City to the Pacific Ocean?

  _____

3.   Mexico City is in map section _____ and is
  (letter-number)
  _____ miles _____ of Hermosillo.
  (number)   (direction)

4.   Tijuana is in map section _____ and is _____
  (letter-number)   (number)
  miles _____ of Mazatlan.
  (direction)

  _____

TYPES OF QUESTIONS

In this next section, we'll work with each of the three types of maps. First, though, we'll look at the three types of questions you'll be asked in your study of maps. These questions will help you find and interpret information on each map.

1) Scanning the Map Questions

"Scanning the Map" questions require you to be familiar with general features concerning correct use of the map. Answer these questions by filling in words to complete a sentence. To scan a map:

- Pay attention to the title.
- Look for any map symbols, distance scales, index guides, special keys, and any other information located on or around the outside of the map.
- Be familiar with boundary lines and other special names and symbols appearing on the map itself.

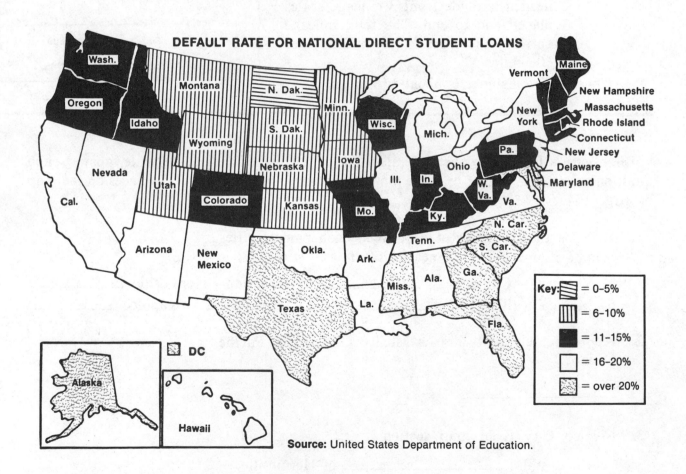

DEFAULT RATE FOR NATIONAL DIRECT STUDENT LOANS

Key:
- = 0-5%
- = 6-10%
- = 11-15%
- = 16-20%
- = over 20%

Source: United States Department of Education.

EXAMPLE 1: The above map shows the state default rate for

_____.

Answer: **National Direct Student Loans** (found in the title)

EXAMPLE 2: The maximum default rate shown in any state is _____.

Answer: **over 20%** (found by looking at the listing of symbols at the right of the map)

2) Reading the Map Questions

"Reading the Map" questions require you to locate and interpret information on the map itself. These questions are answered "true" or "false."

EXAMPLE 1: On loans issued to students, Montana has a default rate somewhere between 6% and 10%.   True   False

Answer: **True.** Montana is shaded with vertical lines. The vertical line symbol represents a rate of 6% to 10%.

EXAMPLE 2: Two states have a default rate in the 0-5% category.   True   False

Answer: **False.** The symbol for 0-5% is horizontal lines. Only 1 state, North Dakota, displays this symbol.

3) Comprehension Questions

"Comprehension" questions require you to compare values, make inferences, and draw conclusions. Answer each question by choosing the best of five possible answers.

EXAMPLE 1: Oregon has a default rate that is _____ Utah's.

a) the same as
b) less than
c) greater than
d) four times
e) one-half

Answer: **c) greater than.** Oregon has a default rate of 11-15%, which is greater than Utah's rate of 6-10%, but not four times greater (choice d).

EXAMPLE 2: The number of states that have a default of over 20% is _____.

a) 5
b) 7
c) 9
d) 10
e) 16

Answer: **c) 9.** Count only the states marked with the shading ▨ . These states are: Alaska, Texas, Mississippi, Florida, Georgia, South Carolina, North Carolina, Maryland, and Delaware. Be careful not to count D.C. (Washington, D.C.), the nation's capital. The question asks only for states.

GEOGRAPHICAL MAPS

A geographical map shows land and water formations. Often, but not always, boundary lines (borders) of countries or states and the locations of cities are also shown. Below are examples of the most commonly used geographical map symbols.

Boundary Lines: show the borders or limits of an area.

Rivers: show large natural streams of water.

Cities: show population centers.

Mountains: show mountainous land formations.

Lakes: show bodies of water surrounded by land.

State Capitals: special symbols show state capitals.

On many geographical maps, color is used to distinguish areas of different height. Land height (elevation) means "distance above sea level." *Sea level* is the level of the ocean. Thus, land that is the same level as the ocean is said to have "0" elevation, while land that is 500 feet higher than the ocean is said to have an elevation of 500 feet.

The map below is an example of a geographical map. Look carefully at the use of symbols on this map before proceeding to the next page.

SOUTHEASTERN UNITED STATES

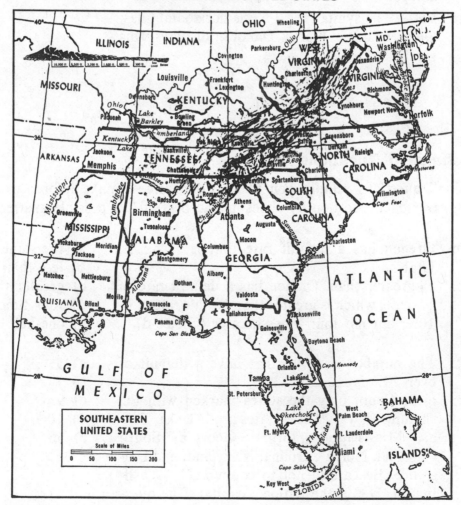

Source: C. S. Hammond, & Co., Maplewood, New Jersey. Used with permission.

To answer questions about geographical maps, follow the sequence below.

SCANNING THE MAP

To scan a geographical map, identify the title, symbols, elevation key, distance scale, and longitude and latitude lines. Not all maps will have all of these items, but you should be aware of what you are working with.

SAMPLE FILL-IN QUESTION:
>The map on the previous page includes the region called the _____.

Answer: Southeastern United States. (found in the title)

READING THE MAP

To read a geographical map, find the place or area asked about. Interpret symbols and shading, and identify longitude and latitude as needed.

SAMPLE TRUE-OR-FALSE QUESTION:
>(Circle your answer.)

Montgomery is the capital of Mississippi.	True False

Step 1.   Look at the symbols for cities, and decide which symbol stands for a state capital. Remember, there is only one state capital in a state.

Step 2.   Find the capital of Mississippi.

Answer: **False.** The capital of Mississippi is Jackson. Montgomery is the capital of Alabama.

COMPREHENSION QUESTIONS

Comprehension questions require you to compare values, make inferences, or draw conclusions.

SAMPLE MULTIPLE-CHOICE QUESTION:
>(Circle the letter of your answer.)

The state that would be the most likely choice for a mountain vacation is _____.	a) Mississippi b) Alabama c) North Carolina d) South Carolina e) Florida

Step 1.   Using the map symbol for mountainous areas, look at each of the states listed at the right.

Step 2.   Decide which state contains the largest amount of mountainous terrain.

Answer: **c) North Carolina**

PRACTICE GEOGRAPHICAL MAP

Source: C. S. Hammond, & Co., Maplewood, New Jersey. Used with permission.

Scanning The Map

Fill in each blank as indicated.

1. The dotted line (----) between 60° and 80° latitude stands for a special latitude line called the
_____

2. The map distance scale represents a total distance of
   _____ miles.

3. The two largest countries on the continent of North
   America are _____ and _____.

4. The _____ degree longitude line passes through the
   center of Cuba.

Reading The Map

Decide whether each sentence is true or false and circle your answer.

5. The United States is north of the 20° latitude line.   True   False

6. Hudson Bay is located in Canada.   True   False

7. The equator passes through the continent of North   True   False
   America.

Comprehension Questions

Circle the letter of the answer that best completes each sentence.

8. You can conclude from the map that the United States
   has _____ land area than Mexico.
   a) less
   b) less populated
   c) more
   d) more populated
   e) none of the above

9. The distance from Los Angeles, California, to New York
   City is approximately _____ miles.
   a) 1,500
   b) 2,000
   c) 2,400
   d) 3,200
   e) 5,000

10. From the map of North America, you can conclude
    that:
    a) North America is characterized by a uniform distribu-
       tion of mountainous areas.
    b) All of the major rivers of North America eventually
       drain into the Gulf of Mexico.
    c) North America contains more natural resources than
       the continent of South America.
    d) North America is the largest of all of the continents
       north of the equator.
    e) North America is characterized primarily by many
       high mountainous areas in the West and by a few low
       mountain ranges in the lowlands of the East.

APPLYING YOUR SKILLS: GEOGRAPHICAL MAPS

CENTRAL AMERICA

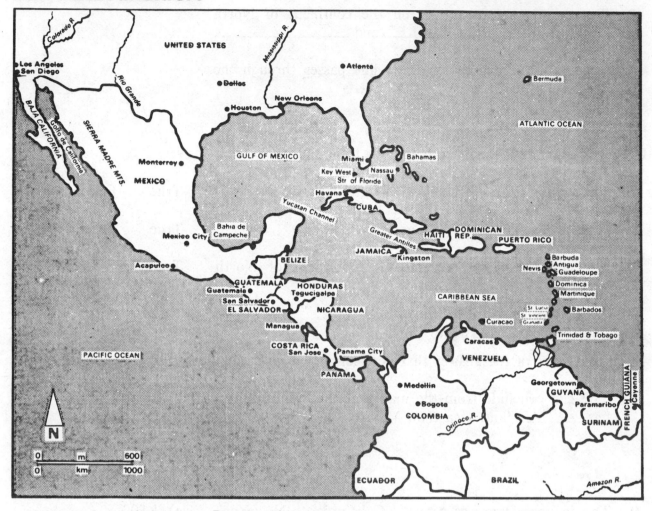

© 1981 Martin Greenwald Assoc., Inc. Reproduced from *Maps on File* with permission of Facts on File, Inc.

Directions: Answer each question by completing the sentence, choosing true or false, or selecting the correct multiple-choice response.

1. El Salvador shares borders with the two countries:
   _____ and _____.

2. Mexico shares borders with two countries of Central America: _____ and _____.

3. The largest island off the southern coast of Florida is _____.

4. Puerto Rico is located _____ of the
   (direction)
   Dominican Republic.

5. The countries of Haiti and the Dominican Republic are part of the same island.

    True   False

6. Havana, Cuba is about _____ miles from Miami, Florida.

    a) 50
    b) 250
    c) 450
    d) 650
    e) 850

7. Cuba is approximately _____ miles from Nicaragua.

    a) 300
    b) 600
    c) 900
    d) 1,200
    e) 1,500

8. Honduras shares borders with the countries of Guatemala, _____ and _____.

    a) Belize, El Salvador
    b) Belize, Nicaragua
    c) El Salvador, Nicaragua
    d) El Salvador, Panama
    e) Panama, Nicaragua

9. To make an overland trip from Nicaragua to El Salvador requires going through the country of _____.

    a) Belize
    b) Costa Rica
    c) Guatemala
    d) Panama
    e) Honduras

10. Looking carefully at the map, you can conclude that:
    a) Central American countries are characterized by a uniform distribution of mountainous areas.
    b) Central American countries are characterized by mountainous interior regions and narrow lowland coastal regions.
    c) Central American countries are mainly lowland river plains.
    d) The Pacific coast regions of Central America contain most of the larger population centers.
    e) The fastest travel route is from north to south.

DIRECTIONAL MAPS

You are probably very familiar with directional maps. City maps are used to locate streets, buildings of interest, and points of interest such as parks. State and country maps are used to show the location of cities and of the highway systems that connect them. Before we work with directional maps, let's look at some commonly used symbols that appear on them.

Roads and highways are shown by thick lines and by thin lines.

⁓ : 2-lane roads

▬ : Divided highway, 4 lanes or more. (Usually an interstate or U.S. highway)

═ :

Highway route numbers are shown as numbers inside special symbols.

⑦⑤ : Interstate highway 75

⑤⑨ : U.S. highway 59

⑫ : State highway 12

Cities are shown by dots and circles.

● , ○ : Cities

✪ , ⊙ : State Capital

Other symbols

✈ : Airport

⤨ : Intersection of roads and major highways

The map below is a section of a directional map. Refer to this map as you read the example questions on the next page.

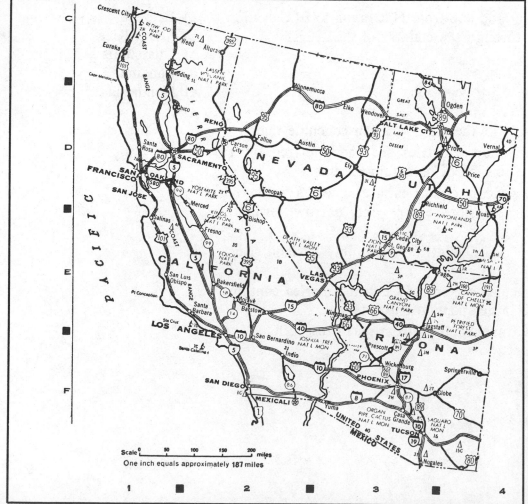

©Rand McNally & Co., R. L. 83-5-59.

Index Guide

Scale
0   50   100   150   200 miles
One inch equals approximately 187 miles

To answer questions about directional maps, follow the sequence below.

SCANNING THE MAP

To scan a directional map, notice the map title, index guide, direction key, and distance scale.

SAMPLE FILL-IN QUESTION:

> According to the distance scale, a map distance of 1 inch represents a real distance of _____ miles.

Answer: **187** (found on the distance scale).

READING THE MAP

To read a directional map, use the index guide to locate places, and use the direction key and distance scale to identify directions and compute distances.

SAMPLE TRUE-OR-FALSE QUESTION:

> (Circle your answer.)
> The major interstate highway running between Phoenix and Los Angeles is ⑰ .   True   False

Step 1. Use the index guide to find both cities.

Step 2. Identify the interstate highway running between them.

Answer: **False.** ⑩ runs between Phoenix and Los Angeles. Notice that ⑰ runs north and south out of Phoenix.

COMPREHENSION QUESTIONS

To answer comprehension questions about directional maps, compare locations and distances, make inferences, and draw conclusions.

SAMPLE MULTIPLE-CHOICE QUESTION:

> (Circle the letter of your answer.)
> The direct distance between Sacramento and Los Angeles is about _____ the direct distance between Sacramento and Reno.

a) 250 miles less than
b) 100 miles less than
c) the same as
d) 250 miles more than
e) 100 miles more than

Step 1. Measure the approximate distance between Sacramento and each other city.
> Sacramento to Los Angeles = 350 miles.
> Sacramento to Reno = about 100 miles.

Step 2. Subtract the distances.
> 350 − 100 = 250

Answer: **d) 250 miles more than**

PRACTICE DIRECTIONAL MAP

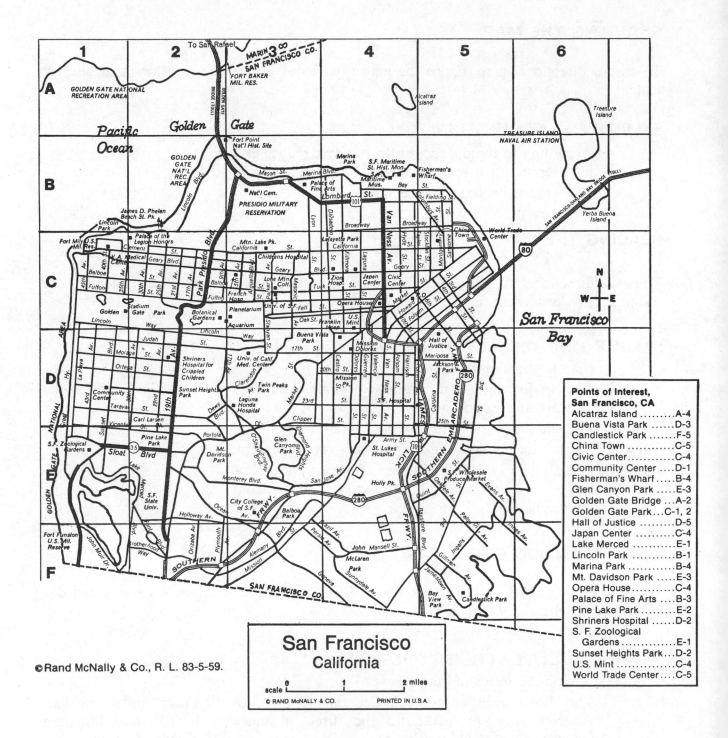

Scanning The Map

Fill in each blank as indicated.

1.   The above map shows points of interest and street names
     in the city of _____.

2. According to the index guide, Alcatraz Island is located in the map section _____.
(letter-number)

3. San Francisco Bay surrounds San Francisco on both the north and _____ sides.
(direction)

Reading The Map

Decide whether each sentence is true or false and circle your answer.

4. Interstate ⑱⓪ is also called the Southern Embarcadero Freeway where it passes through the city of San Francisco.   True   False

5. Candlestick Park is more than 1 mile from San Francisco Bay.   True   False

6. The Japan Center is between Laguna Street and Van Ness Avenue.   True   False

7. Another name for 19th Avenue is Park Presido Blvd.   True   False

Comprehension Questions

Circle the letter of the answer that best completes each sentence.

8. The Botanical Gardens (C-2) are located in _____.

a) Buena Vista Park
b) Glen Canyon Park
c) Golden Gate Park
d) Candlestick Park
e) Mt. Davidson Park

9. The walking distance from the Japan Center to Fisherman's Wharf is about _____.

a) 1 mile
b) 2 miles
c) 3 miles
d) 4 miles
e) 5 miles

10. The directional map of San Francisco contains the following information:
a) location of streets and points of interest
b) street names and major highway numbers
c) land and water elevation
d) b and c
e) a and b

APPLYING YOUR SKILLS: DIRECTIONAL MAPS

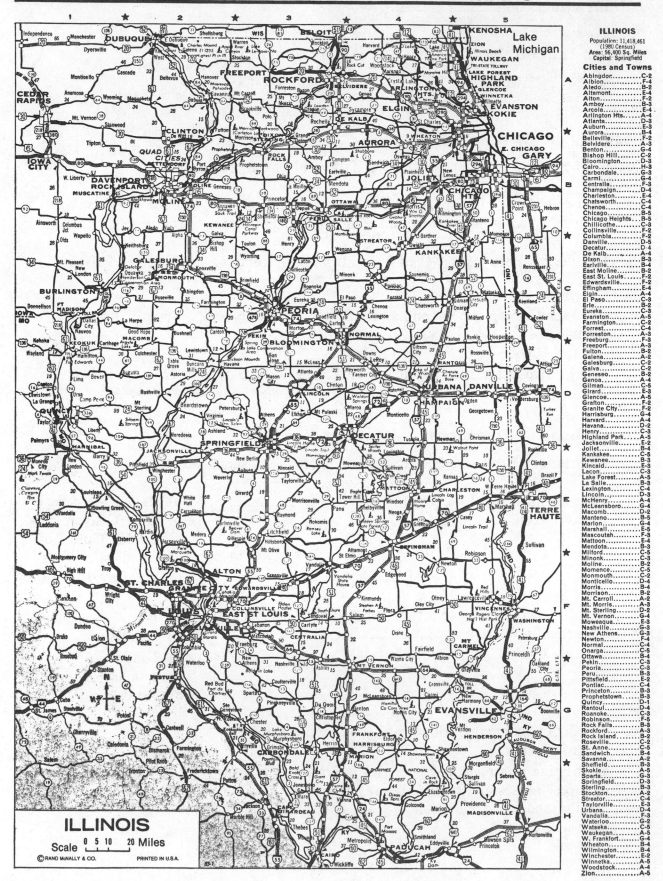

ILLINOIS
Scale 0 5 10 20 Miles
© RAND McNALLY & CO.   PRINTED IN U.S.A.

ILLINOIS
Population: 11,418,461
(1980 Census)
Area: 56,400 Sq. Miles
Capital: Springfield

Cities and Towns

Abingdon............C-2
Albion...............F-4
Aledo...............B-2
Altamont............E-4
Alton...............F-2
Amboy...............B-3
Arcola..............E-4
Arlington Hts........A-4
Atlanta..............D-3
Auburn..............E-3
Aurora..............B-4
Belleville...........F-2
Belvidere............A-3
Benton..............G-4
Bishop Hill..........C-2
Bloomington.........D-3
Cairo...............H-3
Carbondale..........G-3
Carmi...............G-4
Centralia............F-3
Champaign...........D-4
Charleston..........E-4
Chatsworth..........C-4
Chenoa..............C-4
Chicago.............B-5
Chicago Heights......B-5
Chillicothe..........C-3
Collinsville.........F-2
Columbia............F-2
Danville.............D-5
Decatur.............D-4
De Kalb..............A-4
Dixon...............B-3
Earlville.............B-4
East Moline..........B-2
East St. Louis........F-2
Edwardsville.........F-2
Effingham............E-4
Elgin................C-4
El Paso..............C-3
Erie.................B-2
Eureka...............C-3
Evanston.............A-5
Farmington..........C-2
Forrest..............C-4
Forreston............A-3
Freeburg.............F-3
Freeport.............A-2
Fulton...............B-2
Galena...............A-2
Galesburg............C-2
Galva...............C-2
Geneseo.............B-2
Genoa...............A-4
Gilman..............C-5
Girard...............E-3
Glencoe.............A-5
Grafton..............F-2
Granite City.........F-2
Harrisburg...........G-4
Harvard..............A-4
Havana...............D-2
Henry...............C-3
Highland Park........A-5
Jacksonville.........E-2
Joliet...............B-4
Kankakee............C-5
Kewanee.............B-3
Kincaid..............E-3
Lacon...............C-3
Lake Forest..........A-5
La Salle.............B-3
Lexington............C-4
Lincoln..............D-3
McHenry.............A-4
McLeansboro.........G-4
Macomb..............D-2
Manteno.............B-5
Marion..............G-4
Marshall.............E-5
Mascoutah...........F-3
Mattoon.............E-4
Mendota.............B-3
Milford..............C-5
Minonk..............C-3
Moline...............B-2
Momence.............C-5
Monmouth...........C-2
Monticello...........D-4
Morris...............B-4
Morrison.............B-2
Mt. Carroll...........A-2
Mt. Morris...........A-3
Mt. Sterling.........D-2
Mt. Vernon..........G-4
Moweaqua...........E-3
Nashville............G-3
New Athens..........G-3
Newton..............F-4
Normal..............D-3
Onarga..............C-5
Ottawa..............B-4
Pekin................C-3
Peoria...............C-3
Peru................B-3
Pittsfield............E-2
Pontiac..............C-4
Princeton............B-3
Prophetstown........B-3
Quincy..............D-1
Rantoul.............D-4
Roanoke.............C-3
Robinson............F-5
Rock Falls...........B-3
Rockford............A-3
Rock Island.........B-2
Roseville............C-2
St. Anne............C-5
Sandwich............B-4
Savanna.............A-2
Sheffield............B-3
Skokie..............A-5
Sparta..............G-3
Springfield..........D-3
Sterling.............B-3
Stockton.............A-2
Streator.............C-3
Taylorville...........E-3
Urbana..............D-4
Vandalia.............F-3
Waterloo............F-2
Watseka.............C-5
Waukegan...........A-5
W. Frankfort........G-4
Wheaton............B-4
Wilmington..........B-4
Winchester..........E-2
Winnetka............A-5
Woodstock..........A-4
Zion................A-5

© Rand McNally & Co., R. L. 83-5-59.

Directions: Answer each question by completing the sentence, choosing true or false, or selecting the correct multiple-choice response.

1. According to the index guide, the city of Peru is located in the map section _____.
   (letter-number)

2. The largest city in Illinois is _____.

3. The large body of water that Chicago is located near is
   _____.

4. Decatur is about 40 miles west of Springfield.   True   False

5. The direct distance by air between Peoria and Chicago is about 125 miles.   True   False

6. The major interstate approaching Chicago from the south is 94 .   True   False

7. The driving distance between East St. Louis and Chicago is just a little less than _____.
   a) 100 miles
   b) 200 miles
   c) 300 miles
   d) 400 miles
   e) 500 miles

8. If you leave Chicago driving south on 57 , and then turn east on 74 , you will come to the city of _____.
   a) Danville
   b) Normal
   c) Peoria
   d) East St. Louis
   e) Rockford

9. Urbana is about the same distance from Chicago as it is from the city of _____.
   a) Carbondale
   b) Decatur
   c) East St. Louis
   d) Rockford
   e) Galesburg

10. At 55 miles per hour, what is the approximate driving time between Urbana and Chicago?
    Remember: To estimate driving time, divide the distance traveled by the average speed.
    a) $\frac{1}{2}$ hour
    b) 1 hour
    c) 2 hours
    d) $3\frac{1}{2}$ hours
    e) 5 hours

INFORMATIONAL MAPS

An informational map is the most common type of map that appears in newspapers and magazines. Unlike geographical and directional maps, each informational map shows a special kind of information. Examples of informational maps include weather maps, zip code maps, time zone maps, and maps dealing with such concerns as unemployment. We'll work with several of these examples in this section.

One characteristic of informational maps is that each may have its own set of unique symbols. As an example, look at the weather "forecast" map below.

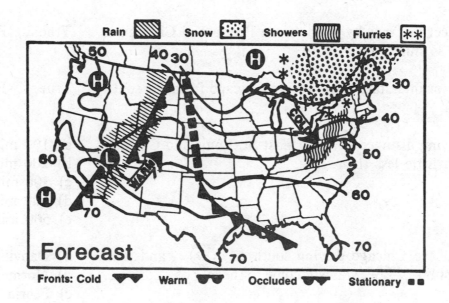

Several symbols are defined on the map, and several we define below.

Symbols Defined on the Map

Snow ▒

Flurries: light snow and wind ✳✳

Rain ▨

Showers: brief periods of rain ⦀

Definitions of Other Symbols

Low atmospheric pressure. A "low" indicates moist air and cool, wet weather. **L**

High atmospheric pressure. A "high" indicates dry air and warm, dry weather. **H**

Note: Temperatures are indicated by numbers around the border of the U.S. The lines connecting the "same temperature" readings indicate locations with the same temperature. These lines are called "constant temperature lines" or *isotherms*.

To answer questions about informational maps, follow the sequence below.

SCANNING THE MAP

To scan an informational map, notice the graph title and any special symbols used on the map.

SAMPLE FILL-IN QUESTION:
The symbol ⊞ means _____.

Answer: Snow flurries

READING THE MAP

To read an informational map, identify map symbols and other data on the map itself. Then, locate the specific information needed.

SAMPLE TRUE-OR-FALSE QUESTION:
(Circle your answer.)
The high temperatures shown for most of the south-   True   False
eastern United States are in the 40's.

Step 1. Locate the southeastern states in the lower right hand corner of the country.

Step 2. Read the temperatures indicated.

Answer: **False.** The high temperatures in the southeastern states are in the 50's, 60's and, to a small extent, the 70's.

COMPREHENSION QUESTIONS

To answer comprehension questions, compare the use of map symbols and other data on the map.

SAMPLE MULTIPLE-CHOICE QUESTION:
(Circle the letter of your answer.)
This could be a weather map during the month of
_____.

a) January
b) May
c) June
d) July
e) August

Step 1. Notice that the moisture shown includes snow and flurries.

Step 2. Combine this with the information that the temperatures range from the 30's to the 70's.

Answer: **a)** January is the best of all the possible answers.

PRACTICE INFORMATIONAL MAP

MILITARY MANPOWER: AFRICA

Military Manpower

South Africa 404,500*
Egypt 395,000
Ethiopia 221,600
Nigeria 173,000
Morocco 98,000
Algeria 78,800
Sudan 62,900
Tanzania 51,700
Somali Democratic Republic 46,500
Angola 40,000
Libya 37,000
Tunisia 34,300†
Mozambique 24,000
Zimbabwe-Rhodesia 21,500
Uganda 21,000

Zaire 20,500
Ghana 20,000
Zambia 14,300
Kenya 12,400
Madagascar 10,500
Mauritania 9,450
Guinea 8,850
Cameroon 8,500
Senegal 8,350
Congo 7,000
Guinea-Bissau 6,100†
Liberia 5,250
Chad 5,200
Burundi 5,000†
Malawi 5,000†

Ivory Coast 4,950
Mali 4,450†
Upper Volta 4,070†
Rwanda 3,750
Togo 3,250†
Djibouti 3,000
Sierra Leone 3,000
Benin 2,200
Niger 2,150
Gabon 1,300
Central African Empire 1,200
Botswana 1,000†

* Total mobilization
† Includes conscripts

POPULATION: AFRICA

Population as of January 1980

Nigeria 75,840,000
Egypt 41,502,000
Ethiopia 32,184,000
Zaire 28,504,000
South Africa 28,096,000
Morocco 20,667,000
Algeria 18,542,000
Sudan 18,378,000
Tanzania 17,643,000
Kenya 15,660,000
Uganda 13,457,000
Ghana 11,936,000
Mozambique 10,172,000
Madagascar 8,461,000

Cameroon 8,250,000
Ivory Coast 7,898,000
Zimbabwe 7,346,000
Upper Volta 6,738,000
Angola (including
  Cabinda) 6,580,000
Mali 6,553,000
Tunisia 6,392,000
Malawi 5,951,000
Zambia 5,740,000
Senegal 5,591,000
Niger 5,425,000
Guinea 5,351,000
Rwanda 4,639,000

Chad 4,574,000
Burundi 4,366,000
Somalia 3,510,000
Benin 3,427,000
Sierra Leone 3,391,000
Libya 2,933,000
Togo 2,564,000
Central African
  Republic 2,446,000
Liberia 1,818,000
Mauritania 1,574,000
Peoples Republic of the
  Congo 1,525,000
Lesotho 1,321,000

Namibia 1,007,000
Mauritius 943,000
Botswana 780,000
Guinea-Bissau 640,000
The Gambia 593,000
Gabon 585,000
Swaziland 542,000
Réunion 507,000
Equatorial Guinea 345,000
Cape Verde 332,000
Comoros 327,000
Djibouti 318,000
São Tome & Principe 83,000
Seychelles 65,000

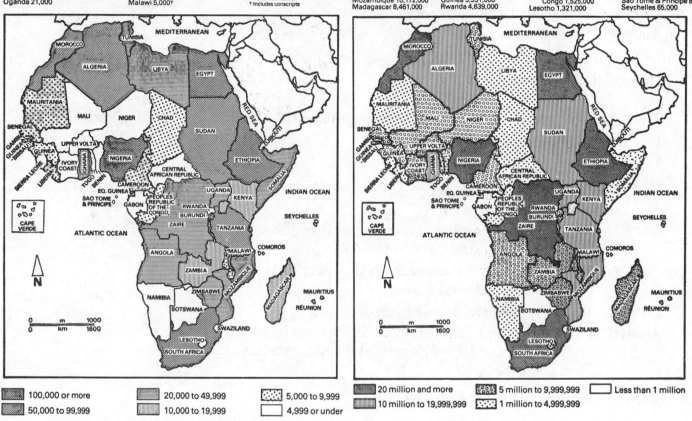

100,000 or more	20,000 to 49,999	5,000 to 9,999	20 million and more 5 million to 9,999,999 Less than 1 million
50,000 to 99,999	10,000 to 19,999	4,999 or under	10 million to 19,999,999 1 million to 4,999,999

Scanning The Maps

Fill in each blank as indicated.

1. The two maps pictured above show the _____ _____ and the _____ _____ of the countries of the continent of Africa.

2. There are _____ countries in Africa with a population of 20 million or more.

3. The population map represents the population as it was as of _____.

4. The symbol ▦ stands for _____ _____ on the population map and for _____ _____on the military manpower map.

Reading The Maps

Decide whether each sentence is true or false and circle your answer.

5. The African country with the largest population also has the greatest number of people in the military.   True   False

6. In Egypt, approximately one person in one hundred is in the military.   True   False

7. Lesotho is completely surrounded by the nation of South Africa.   True   False

Comprehension Questions

Circle the letter of the answer that best completes each sentence.

8. In 1980, the population of Chad was about _____ the population of Libya.
   a) one half
   b) $1\frac{1}{2}$ times
   c) 2 times
   d) $2\frac{1}{2}$ times
   e) 3 times

9. Three island nations with a population of less than a million people are Seychelles, Comoros, and _____.
   a) Madagascar
   b) Equitorial Guinea
   c) Swaziland
   d) Cape Verde
   e) Senegal

10. All of the following statements can be concluded from the two maps EXCEPT:
    a) Egypt is the second most populated country in Africa and also has the second largest military.
    b) Nigeria has almost twice Egypt's population, while Egypt has more than twice Nigeria's military manpower.
    c) The four most populated countries in Africa are also the four with the largest militaries.
    d) Three of the four most populated countries in Africa share a common border with Sudan.
    e) Zaire in south central Africa shares borders with nine other African countries.

APPLYING YOUR SKILLS: INFORMATIONAL MAPS

HEATING YOUR HOME: WHICH FUEL IS MOST ECONOMICAL?

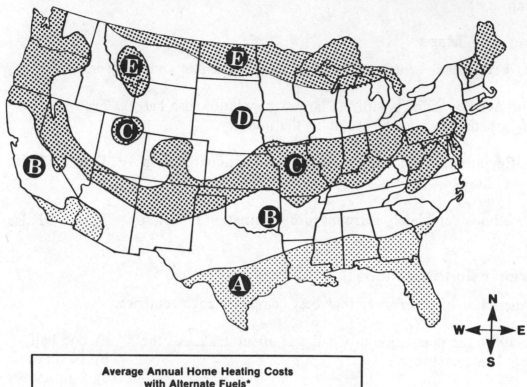

Average Annual Home Heating Costs with Alternate Fuels*					
	Zone				
Fuel	A	B	C	D	E
Oil	$250	$625	$875	$1,050	$1,250
Gas	175	440	500	685	910
Electricity	400	990	1,390	1,680	1,975
Coal	175	315	445	540	640
Wood	150	345	520	620	730

*Based on a 1,000-square-foot home with average insulation and weatherization, 1981 fuel prices. Local climate and fuel price difference will cause heating costs to vary.

Directions: Answer each question by completing the sentence, choosing true or false, or selecting the correct multiple-choice response.

1. The table and map above compare the average annual costs of _____ for each of _____ zones or areas of the country.
    (number)

2. In every zone, the cost of _____ is higher than any other listed fuel.

3. The zone showing the lowest average cost for each of the 5 alternative fuels is _____.

4. Residents of Zone B who heat their home with gas can   True   False
    expect an average cost of $625 per year.

5. Residents of Zone B who use oil pay higher heating costs than residents of Zone C who use wood.

True   False

6. The only zone in which it costs less to heat a home with wood than with coal is _____.

a) Zone A
b) Zone B
c) Zone C
d) Zone D
e) Zone E

7. According to the table and map, states in Zone D pay four times as much for electricity as states in _____.

a) Zone A
b) Zone B
c) Zone C
d) Zone D
e) Zone E

8. The cost per year of heating with gas in Zone B is approximately _____ less than in Zone E.

a) $350
b) $470
c) $530
d) $245
e) $140

9. In Zone D, the cost per year of heating with electricity is _____ more than the cost of heating with wood.

a) $460
b) $760
c) $860
d) $960
e) $1,060

10. The statement that best summarizes the information on this table and map is:
   a) Heating costs are consistently higher in states on the West Coast than in states on the East Coast.
   b) Electricity will continue to be the most expensive source of heating costs in future years.
   c) Oil and gas, used as heating fuels, are more expensive in southern states than in northern states.
   d) Although costs vary from zone to zone, heating costs are consistently lower in southern states than in northern states.
   e) Many people are changing to electric heat from coal because electricity is an environmentally cleaner source of energy.

FINAL MAP SKILLS INVENTORY

To find out how well you understand maps, answer the questions in this inventory. Answer carefully, but do not use outside help. When you've finished, check your answers on page 165.

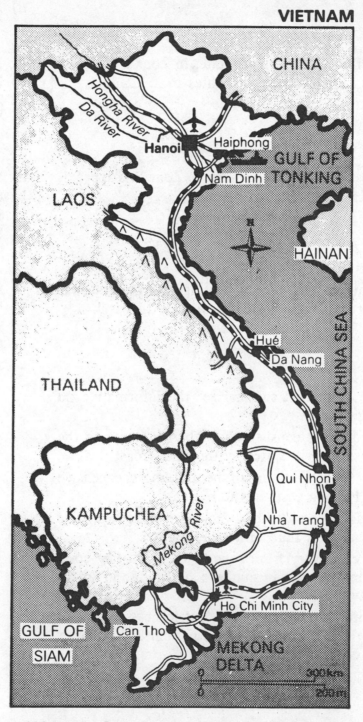

Directions: Answer each question by completing the sentence, choosing true or false, or selecting the best multiple-choice answer.

1. The distance scale measures 200 miles and _____ kilometers.

2. The _____ River flows through Kampuchea.

3. Ho Chi Minh City is located in southern Laos.   True   False

4. The island of Hainan lies approximately 200 miles off the coast of Vietnam.   True   False

5. The two cities in which one can find an airport are _____.
   a) Can Tho and Qui Nhon.
   b) Ho Chi Minh City and Hanoi.
   c) Hanoi and Haiphong.
   d) Can Tho and Hanoi.
   e) Nam Dinh and Ho Chi Minh City.

6. Laos and Vietnam are separated by _____.
   a) the Gulf of Tonking.
   b) the Hongha River.
   c) the Hanoi airport.
   d) a mountain range.
   e) Thailand.

7. The distance from Can Tho to Ho Chi Minh City is approximately _____. (NOTE: km = kilometer)
   a) 150 kilometers.
   b) 200 miles.
   c) 600 kilometers.
   d) 3 kilometers.
   e) 10 miles.

8. The road between Haiphong harbor and Nam Dinh:
   a) is longer than the Da River.
   b) crosses a mountain range.
   c) is approximately the same length as the road between Hue and Da Nang.
   d) crosses the Hongha River.
   e) both c and d above.

MAP B

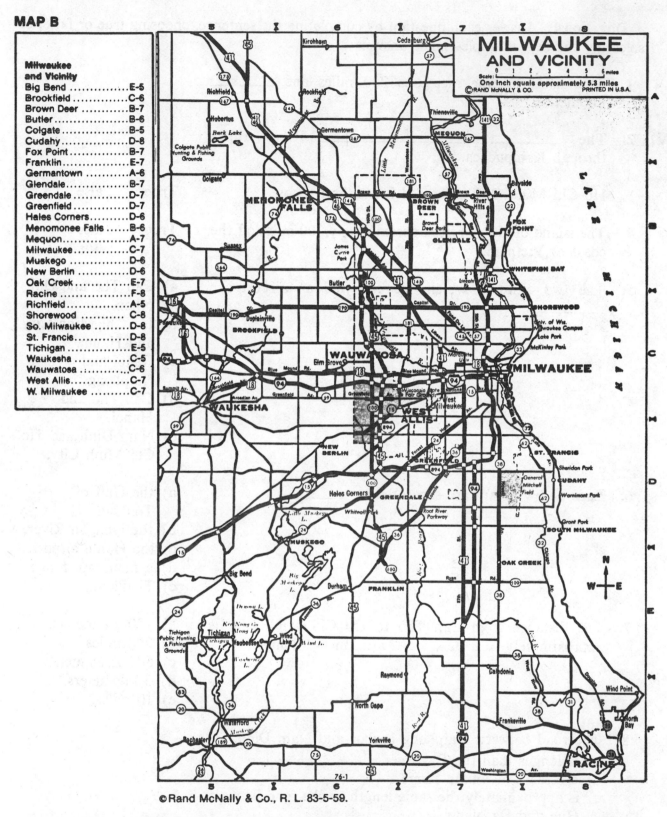

Milwaukee and Vicinity	
Big Bend	E-5
Brookfield	C-6
Brown Deer	B-7
Butler	B-6
Colgate	B-5
Cudahy	D-8
Fox Point	B-7
Franklin	E-7
Germantown	A-6
Glendale	B-7
Greendale	D-7
Greenfield	D-7
Hales Corners	D-6
Menomonee Falls	B-6
Mequon	A-7
Milwaukee	C-7
Muskego	D-6
New Berlin	D-6
Oak Creek	E-7
Racine	F-8
Richfield	A-5
Shorewood	C-8
So. Milwaukee	D-8
St. Francis	D-8
Tichigan	E-5
Waukesha	C-5
Wauwatosa	C-6
West Allis	C-7
W. Milwaukee	C-7

© Rand McNally & Co., R. L. 83-5-59.

Directions: Answer each question by completing the sentence, choosing true or false, or selecting the best multiple-choice answer.

9. On the distance scale, one inch represents a distance of
_____ miles.
    (number)

10. The index guide indicates that Mequon is located in the
section _____.
(letter-number)

11. Menomonee Falls is located southwest of Milwaukee.   True   False

12. Driving at 55 miles per hour, you would need to plan for   True   False
about $1\frac{1}{2}$ hours of driving time for a trip from Milwaukee
to Chicago (a distance of 90 miles).

13. If you drive west from Milwaukee on (94) , and then   a) Wauwatosa
turn north on (45) , and then turn west on (190) , you   b) Waukesha
will reach the town of _____.   c) Butler
　 d) Duplainville
　 e) Greenfield

14. From this map, you can tell that Milwaukee is located   a) state capital
near a _____.   b) lake
　 c) ocean
　 d) mountain range
　 e) plateau

15. Highway Route (190) is also named _____.   a) Brown Deer Rd.
　 b) Greenfield Ave.
　 c) Capitol Dr.
　 d) National Ave.
　 e) Loomis Rd.

16. Traveling north on Route (41) , the distance between   a) 5.3
the Route (20) turnoff and the Route (100) turnoff is   b) 11
about _____ miles.   c) 50
　 d) 80
　 e) 100

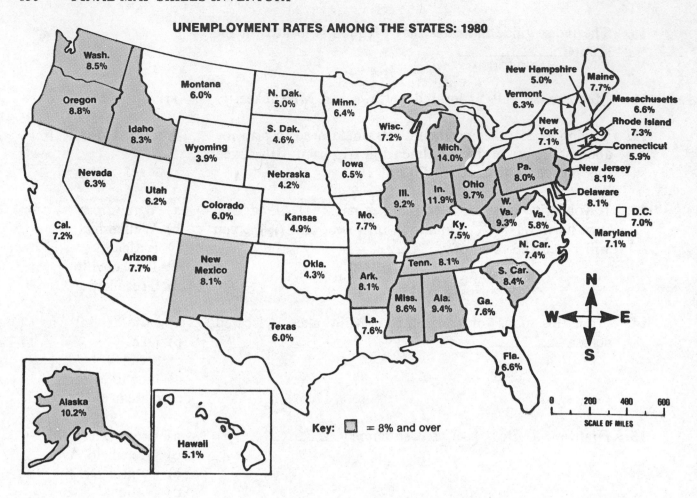

UNEMPLOYMENT RATES AMONG THE STATES: 1980

Key: ▨ = 8% and over

SCALE OF MILES

0   200   400   600

Directions: Answer each question by completing the sentence, choosing true or false, or selecting the best multiple-choice response.

17. The map above shows the percent of _____ _____ for each state in the United States during the year _____.

18. States that are lightly shaded had unemployment rates of _____.

19. In 1980, Florida's unemployment rate was 6.6%.   True   False

20. Idaho's 1980 unemployment rate was less than 8%.   True   False

21. The state shown to have the highest unemployment rate during the year 1980 is _____.

    a) Alabama
    b) Michigan
    c) Kansas
    d) Texas
    e) Alaska

22. The area with the lowest unemployment rates is located in the _____.

    a) northwest states
    b) southwest states
    c) central states
    d) southeast states
    e) northeast states

23. A person leaving the farm state of Kansas could expect to find _____ job opportunities in the coal-producing state of West Virginia.

    a) more
    b) fewer
    c) an equal number of
    d) mainly clerical
    e) mainly manufacturing

24. During 1980, the unemployment rate in Alaska was _____ as large as the unemployment rate in Hawaii.

    a) half
    b) two times
    c) one third
    d) three times
    e) one fourth

FINAL MAP SKILLS INVENTORY CHART

Circle the number of any problem that you missed and be sure to review the appropriate page. A passing score is 19 correct answers. If you miss more than 5 questions, you should review this chapter.

Circle the Questions Missed	Type of Map	Type of Question	Practice Page
1	Geographical	Scanning	109
9, 10	Directional	Scanning	115
17, 18	Informational	Scanning	121
2, 3, 5, 6	Geographical	Reading	109
11, 14, 15	Directional	Reading	115
19, 20	Informational	Reading	121
4, 7, 8	Geographical	Comprehension	109
12, 13, 16	Directional	Comprehension	115
21, 22, 23, 24	Informational	Comprehension	121

BUILDING GRAPH, CHART, AND MAP POWER: REVIEW TEST

This test will show you how well you have learned all of the skills that you have practiced in this section of the book. Take your time and work each problem carefully. Answer the questions by completing the sentence, answering true or false, or selecting the best multiple choice response.

GRAPH A

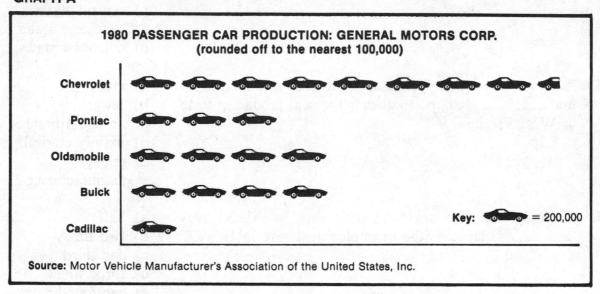

1980 PASSENGER CAR PRODUCTION: GENERAL MOTORS CORP.
(rounded off to the nearest 100,000)

Key: = 200,000

Source: Motor Vehicle Manufacturer's Association of the United States, Inc.

1. Graph A shows the passenger car production by _____, with the symbol represent
   (manufacturer)
   senting _____ cars.

2. In 1980, General Motors Corporation produced 4,200,000 cars.   True   False

3. According to Graph A, for every 1 Cadillac made, _____ Chevrolets were produced.
   a) 1
   b) 7
   c) $7\frac{1}{2}$
   d) $8\frac{1}{2}$
   e) 9

4. The statement that best describes Graph A is:
   a) General Motors Corporation produces more cars than any other company.
   b) Cadillacs sell at a higher price than Buicks.
   c) More Chevrolets are produced than all other makes of GM cars combined.
   d) General Motors expected to sell more Chevrolets than Pontiacs, Oldsmobiles, Buicks, or Cadillacs.
   e) Pontiacs, Oldsmobiles, and Buicks are medium-sized cars.

GRAPH B

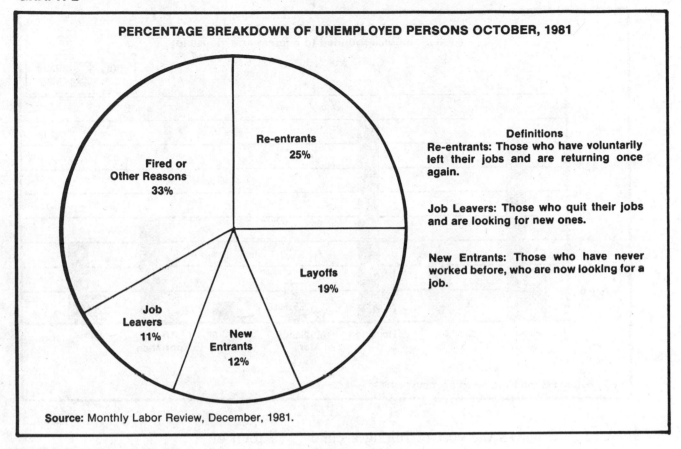

PERCENTAGE BREAKDOWN OF UNEMPLOYED PERSONS OCTOBER, 1981

Re-entrants
25%

Fired or
Other Reasons
33%

Layoffs
19%

Job
Leavers
11%

New
Entrants
12%

Definitions
Re-entrants: Those who have voluntarily left their jobs and are returning once again.

Job Leavers: Those who quit their jobs and are looking for new ones.

New Entrants: Those who have never worked before, who are now looking for a job.

Source: Monthly Labor Review, December, 1981.

5. Graph B shows the percentage breakdown of _____ _____ people during the month of October, 1981.

6. One-quarter of the unemployed are workers who are re-entering the labor force, such as women who have taken time off to bear children.

True   False

7. The total percentage of unemployed who have been laid off is almost _____ than those who are leaving their jobs.

a) one-third less
b) two times greater
c) three times less
d) three times greater
e) one-half less

8. The statement that best describes Graph B is:
a) The largest group of the unemployed was made up of re-entrants.
b) The smallest percentage of the unemployed was made up of new entrants.
c) During October 1981, the categories called "layoffs" and "fired or other reasons" together made up the majority of the unemployed.
d) The majority of those unemployed was composed of job leavers, new entrants, and re-entrants.
e) More people are re-entering the job force than ever before.

GRAPH C

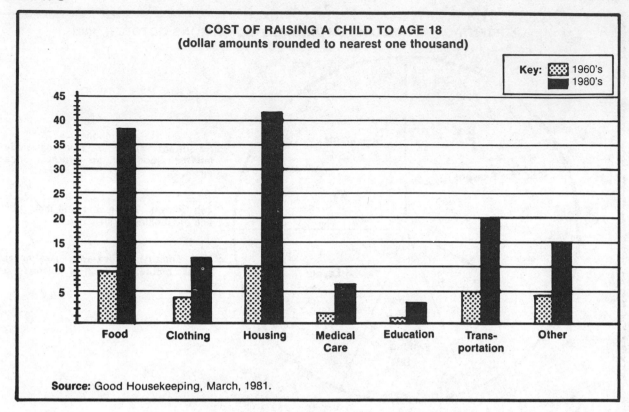

COST OF RAISING A CHILD TO AGE 18
(dollar amounts rounded to nearest one thousand)

Key: 1960's
     1980's

Food   Clothing   Housing   Medical Care   Education   Transportation   Other

Source: Good Housekeeping, March, 1981.

9. Graph C shows the cost of raising a child for a total of
_____ years.
   (number)

10. From the 1960's to the 1980's, the greatest increase was in   True   False
food costs.

11. From the 1960's to the 1980's, the increase in the cost of   a) $35,000
raising a child to age 18 was approximately _____.   b) $50,000
   c) $100,000
   d) $150,000
   e) $138,000

12. All the following statements can be concluded from Graph
C EXCEPT:
   a) The cost of raising a child to age 18 is greater in the
   1980's than in the 1960's.
   b) The cost of raising a child in the 1980's is more than
   three times as expensive as it was in the 1960's.
   c) Housing is the most costly area in both the 1960's and
   the 1980's.
   d) In the 1980's, housing accounts for one-half of all the
   costs.
   e) The cost of food is four times greater in the 1980's
   than in the 1960's.

GRAPH D

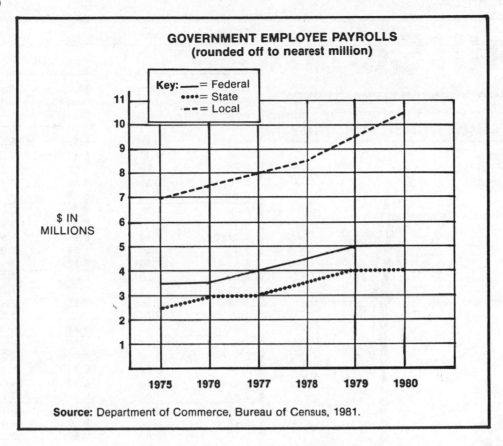

GOVERNMENT EMPLOYEE PAYROLLS
(rounded off to nearest million)

Key: —— = Federal
•••• = State
- - = Local

$ IN MILLIONS

1975   1976   1977   1978   1979   1980

Source: Department of Commerce, Bureau of Census, 1981.

13. Consistently, the highest total payroll has been composed of those at the _____ level.

14. From 1979 to 1980, there was little, if any, increase in the total payroll at the federal level.

    True   False

15. The smallest difference in total payrolls between government employees at the state and federal levels was in the year _____.

    a) 1975
    b) 1976
    c) 1977
    d) 1978
    e) 1979

16. The statement that best describes Graph D is:
    a) Payrolls for state employees are greater than those for federal employees.
    b) Payrolls for governmental employees at all levels have gradually increased between 1975 and 1980.
    c) The greatest gain in employee payrolls has occurred at the federal levels.
    d) a and b
    e) b and c

SCHEDULE A:

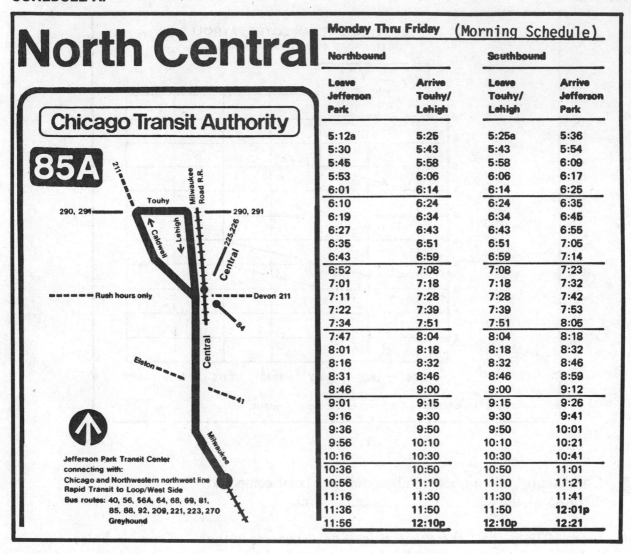

Monday Thru Friday (Morning Schedule)			
Northbound		**Southbound**	
Leave Jefferson Park	Arrive Touhy/ Lehigh	Leave Touhy/ Lehigh	Arrive Jefferson Park
5:12a	5:25	5:25a	5:36
5:30	5:43	5:43	5:54
5:45	5:58	5:58	6:09
5:53	6:06	6:06	6:17
6:01	6:14	6:14	6:25
6:10	6:24	6:24	6:35
6:19	6:34	6:34	6:45
6:27	6:43	6:43	6:55
6:35	6:51	6:51	7:05
6:43	6:59	6:59	7:14
6:52	7:08	7:08	7:23
7:01	7:18	7:18	7:32
7:11	7:28	7:28	7:42
7:22	7:39	7:39	7:53
7:34	7:51	7:51	8:05
7:47	8:04	8:04	8:18
8:01	8:18	8:18	8:32
8:16	8:32	8:32	8:46
8:31	8:46	8:46	8:59
8:46	9:00	9:00	9:12
9:01	9:15	9:15	9:26
9:16	9:30	9:30	9:41
9:36	9:50	9:50	10:01
9:56	10:10	10:10	10:21
10:16	10:30	10:30	10:41
10:36	10:50	10:50	11:01
10:56	11:10	11:10	11:21
11:16	11:30	11:30	11:41
11:36	11:50	11:50	12:01p
11:56	12:10p	12:10p	12:21

North Central

Chicago Transit Authority

85A

211
290, 291
Touhy
Milwaukee Road R.R.
Caldwell
Lehigh
290, 291
225-226
Central
Rush hours only
Devon 211
84
Elston
Central
41
Milwaukee

Jefferson Park Transit Center connecting with:
Chicago and Northwestern northwest line
Rapid Transit to Loop/West Side
Bus routes: 40, 56, 56A, 64, 68, 69, 81,
85, 88, 92, 209, 221, 223, 270
Greyhound

17. Schedule A shows the bus schedule for the route number
     _____, which is named _____.

18. The Northbound bus leaves from Touhy/Lehigh Streets.   True   False

19. According to Schedule A, the time it takes to go from   a) is 13 minutes
     Jefferson Park to Touhy/Lehigh _____.   b) is 14 minutes
                                                                 c) is 15 minutes
                                                                 d) is always the same
                                                                 e) varies with the time
                                                                    of the morning

20. The total time it takes the bus leaving at 10:56 from   a) 20
     Jefferson Park to make a round trip is _____ minutes.   b) 24
                                                                 c) 31
                                                                 d) 25
                                                                 e) 35

CHART A

A NUTRITIVE VALUE OF DAIRY FOODS					
Food Product	Calories	Protein(g)	Fat(g)	Calcium(g)	Vitamin A
Cheese-Cheddar (1 oz.)	115	7	9	204	300 (I.U.)
Cottage Cheese (1 cup)	220	26	9	126	340
Swiss Cheese (1 oz.)	105	8	8	272	240
Whole Milk (1 cup)	150	8	8	291	310
Nonfat Milk (1 cup)	85	8	T	302	500
Ice Cream (1 cup)	270	5	14	176	540
Sherbet (1 cup)	270	2	4	103	190

Source: U.S. Department of Agriculture, *Home and Garden Bulletin No. 72,* 1978

21. Chart A is based on information from the U.S. Dept. of
_____.

22. From Chart A, you can determine the number of calories   True   False
in one cup of Swiss cheese.

23. The difference in calcium between whole milk and nonfat   a) 11
milk is _____ grams per cup.   b) 75
          c) 83
          d) 19
          e) 146

24. Which graph below most accurately includes the information on the above chart?

a) calories

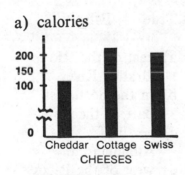

b) protein (g)

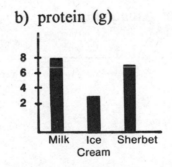

c) fat (g)

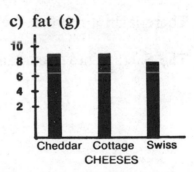

d) calcium (g)

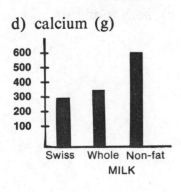

e) vitamin A

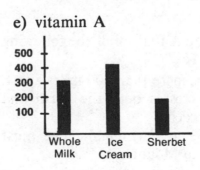

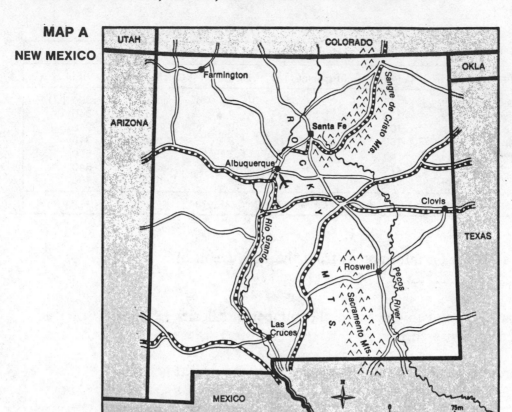

MAP A
NEW MEXICO

25. Map A shows the state of New Mexico, which is bordered by the states of Utah, _____, Oklahoma, Texas, and _____.

26. There is an airport in the city of Albuquerque.   True   False

27. The Sacramento Mountains are located _____.

a) east of the Rio Grande River
b) in the northeast part of the state
c) north of Albuquerque
d) west of the Pecos River
e) both a and d

28. It can be concluded from Map A that all of the following statements are true EXCEPT:

a) Las Cruces and Clovis are more than 100 miles apart.
b) The Rio Grande River crosses both the north and south border of New Mexico.
c) To travel from Sante Fe to Farmington, one must cross the Sangre de Cristo Mountains.
d) The distance from Santa Fe to Albuquerque is less than 75 miles.
e) Farmington is the northernmost city shown on Map A.

MAP B

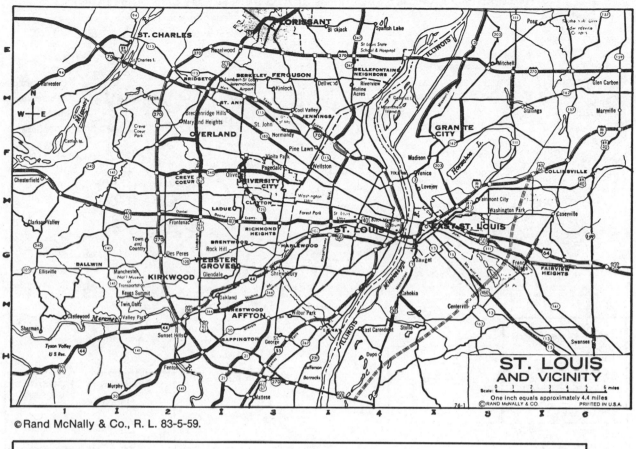

© Rand McNally & Co., R. L. 83-5-59.

St. Louis & Vicinity

Affton....................H-3	Collinsville..............F-6	Granite City F-5	St. Louis............F-3, 4;
Bellefontaine	Crestwood..........H-2, 3	Jennings................F-4	G-3, 4
Neighbors............E-4	Creve Coeur............F-2	Kirkwood................G-2	University City F-3
Berkeley E-3	East St. Louis G-4, 5	Maplewood.............G-3	Webster Grove.........G-3
Brentwood G-3	Fairview Hts. G-6	Overland F-3	
Bridgeton................E-2	Ferguson E-3	Richmond Hts..........G-3	
Clayton F-3	Florissant...............E-3	St. Charles E-1	

29. Webster Groves is located in map section _____.
(letter-number)

30. Granite City is located in Illinois.   True   False

31. Richmond Heights is located:
a) 3 miles northeast of University City.
b) less than 20 miles west of St. Louis.
c) less than 20 miles east of St. Louis.
d) 6 miles south of Florissant.
e) on the border between Missouri and Illinois.

32. Affton is located at the junction of (30) and
_____
a) (40)
b) (270)
c) (21)
d) (13)
e) (15)

MAP C

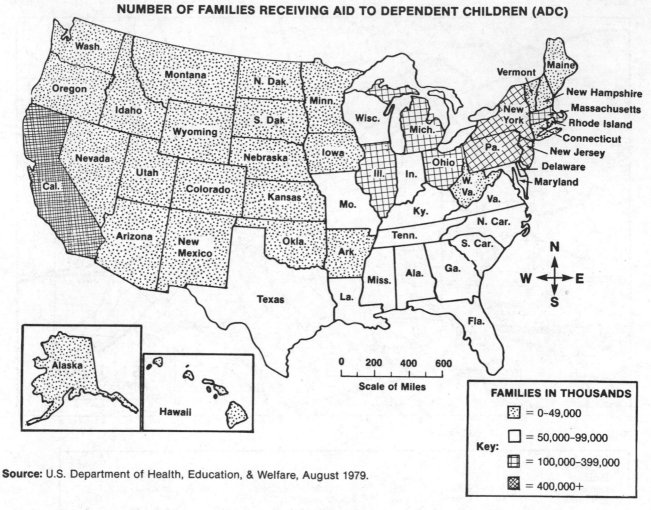

NUMBER OF FAMILIES RECEIVING AID TO DEPENDENT CHILDREN (ADC)

Source: U.S. Department of Health, Education, & Welfare, August 1979.

33. Map C shows the number of families receiving _____.

34. More than 100,000 but less than 400,000 families with   True   False
dependent children receive federal funds in the state of
California.

35. States in which 50,000 to 99,000 families receive assis-   a) northwestern
tance are primarily located in the _____ United States.   b) northern
    c) northeastern
    d) southeastern
    e) southwestern

36. From Map C, it can be concluded that:
a) Most states have fewer than 100,000 families receiving
   assistance.
b) Displaced farm families constitute the majority of
   people receiving aid.
c) California has the highest rate of unemployment.
d) States with higher populations receive more aid.
e) Illinois has 100,000 individuals receiving assistance.

REVIEW TEST INVENTORY CHART

Circle the number of any problem that you miss and be sure to review the appropriate practice page. A passing score is 29 correct answers.

Review any remaining problem areas. If you passed the test, go on to Using Number Power. If you did not pass the test, take the time to make a more thorough review of the book.

Circle the Questions Missed	Type	Type of Question	Practice Pages
1, 5, 9	Graph		10, 22, 34, 46
17, 18, 21	Chart/Schedule	Scanning	72
25, 29, 33	Map		109, 115, 121
2, 6, 13	Graph		11, 23, 35, 47
22	Chart/Schedule	Reading	72
26, 27, 30, 31, 32, 34, 35	Map		109, 115, 121
3, 4, 7, 8, 10, 11, 12, 14, 15, 16	Graph		11, 23, 35, 47
19, 20, 23, 24	Chart/Schedule	Comprehension	72
28, 36	Map		109, 115, 121

Budgeting money is an important financial skill. Successful individuals and companies use budgeting to plan the flow of money.

Rather than spending money in an unplanned way, budgeting allows an individual to carefully plan the percent of income that should go for food, housing, clothing, and other important monthly costs. The circle graph below is an example of family budgeting.

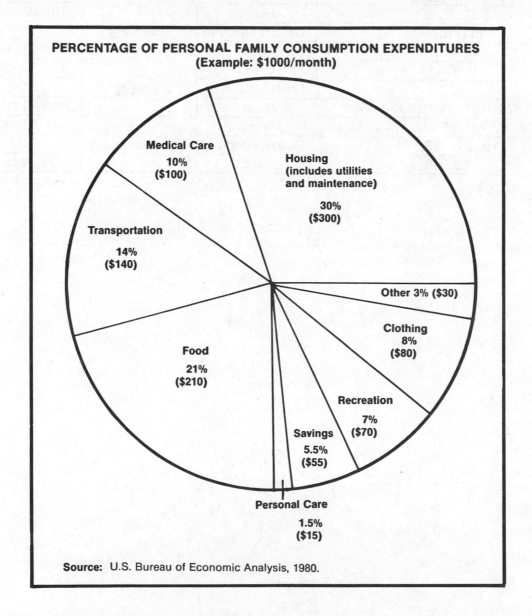

PERCENTAGE OF PERSONAL FAMILY CONSUMPTION EXPENDITURES
(Example: $1000/month)

Medical Care 10% ($100)

Housing (includes utilities and maintenance) 30% ($300)

Transportation 14% ($140)

Other 3% ($30)

Clothing 8% ($80)

Food 21% ($210)

Recreation 7% ($70)

Savings 5.5% ($55)

Personal Care 1.5% ($15)

Source: U.S. Bureau of Economic Analysis, 1980.

Use this graph to answer questions 1-5.

1. According to the graph, the percentages for housing and food costs add up to slightly more than half of the family's total expenditures.   True   False

After housing and food, medical care is the largest budgeted family expenditure.

True   False

If a family's take-home pay was $1,000, the budget allowed the family to spend up to _____ for food.

a) $140
b) $300
c) $210
d) $100
e) $21

Your family did not have any medical expenses this month. You decided to put the money budgeted for medical expenses into savings. What percent of your family's income went into savings for the month?

a) 21.5%
b) 15.5%
c) 25.5%
d) 25%
e) 4.5%

The Smith Family prepared the following budget. Using the information on the graph, circle the items that are incorrectly budgeted and make the necessary corrections. The first two lines have been done for you. Remember, in a "balanced" budget, total expenses equal total income.

Income:
    Take-Home Pay = $1,000.00

Expenses:		Adjustments (if any)	Corrected Budget
Food	$ 210	0	210
Clothing	100	−20	80
Personal Care	18		
Housing	300		
Medical Care	100		
Transportation	145		
Recreation	130		
Savings	55		
Other	40		
Total Expenses:	$1098	$	$

Which phrase best describes the uncorrected budget? **Check one:**
    Over budget (more expenses than income)   ☐
    Balanced budget (expenses equal income)   ☐

COST OF HOUSING: BUYING A HOUSE

Rising housing costs have made it impossible for many people to buy a home. Housing costs have been affected by a number of factors, including high interest rates for loans and increased costs of building materials.

The cost of housing varies widely across the country. It is generally recommended that no more than 25-30% of a family's income should go for housing. However, in some cities, the percentage of income that goes for housing greatly exceeds the recommended maximum percentage. As a result, many families must do without other things in order to make house payments.

	HOUSING COSTS, INCOME*, AND MORTGAGE PAYMENTS			
Cities	Cost of Existing Housing	Average Family Income per Year	Estimated Monthly House Payment	Percentage of Income on House Payments**
San Francisco	$133,900	$31,470	$1,480	66.0%
Washington, D.C.	$100,000	$39,720	$1,005	39.6%
Houston	$ 95,000	$31,810	$ 976	44.0%
Minneapolis	$ 82,300	$33,640	$ 830	39.3%
Chicago	$ 77,200	$31,550	$ 812	38.4%
Atlanta	$ 75,000	$29,430	$ 755	39.5%
Pittsburgh	$ 59,000	$29,790	$ 625	34.1%

Source: *Crain's Chicago Business*, August 1981
* Assuming two incomes per family
** Recommended maximum is 30%

1. The average yearly income for a family in San Francisco is $_____ more per year than for a family in Pittsburgh.

2. Basing your answers on the chart, who has more money to spend after the house payment is made: a family living in Houston or a family living in Atlanta? (Hint: First find monthly income by dividing yearly income by the number of months per year.)

_____

3. Based on the chart, the percentage of income paid for house payments in Pittsburgh is _____ than in San Francisco.

    a) 22.0% more
    b) 26.4% more
    c) 31.9% more
    d) 31.9% less
    e) 27.5% less

COST OF HOUSING: BUYING A HOUSE

4. A conclusion that can be drawn from the chart is:

a) The larger the city, the larger the expected monthly house payment.

b) Average family incomes vary according to the tax burden assumed by the family.

c) In the seven cities listed, the percentage of income spent on housing exceeds the recommended maximum, creating a financial burden on families.

d) The percentage of two-income families is increasing rapidly.

e) The cost of existing housing is expected to decline.

5. Based on the chart below, could you afford the monthly house payments along with other family expenses, if you were. . .   (Circle your answer)

Living in	and Your Monthly* Income is . . .	and Your House + Other Payment is Expenses	Answer	
Washington, D.C.	$3,310	$1,005 + $2,032 ?	Yes	No
Chicago	$2,629	$ 812 + $1,820 ?	Yes	No
Pittsburgh	$2,482	$ 625 + $1,635 ?	Yes	No
San Francisco	$2,622	$1,480 + $1,400 ?	Yes	No
Minneapolis	$2,803	$ 830 + $1,701 ?	Yes	No

*assuming two incomes per family.

COST OF UTILITIES: PAYING YOUR ELECTRIC BILL

One of the major concerns of people throughout the world is the cost of energy. Oil, gas, and electric rates have risen to record highs during the past several years.

Study the information below to determine how electric rates compare from one location in the country to another.

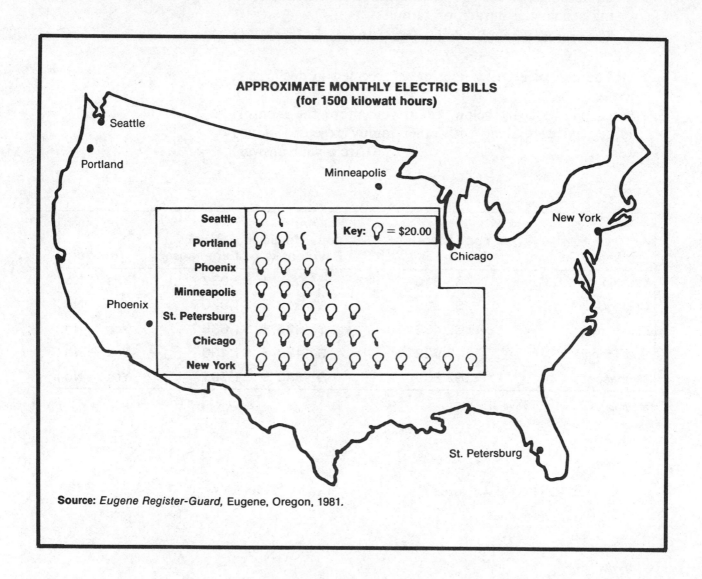

APPROXIMATE MONTHLY ELECTRIC BILLS
(for 1500 kilowatt hours)

Key: 💡 = $20.00

Source: *Eugene Register-Guard*, Eugene, Oregon, 1981.

1. If you were in New York City, the cost of your electric bill would be $_____ more per month than if you were in St. Petersburg.

2. Compute the following electric bills based on the number of kilowatt hours. Remember the chart above is based on a use of 1,500 kilowatt hours.

City	Kilowatt Hours Used	Monthly Electric Bill
Phoenix	750 hours	$
Portland	1,500 hours	$
Chicago	1,500 hours	$
St. Petersburg	3,000 hours	$
New York City	3,000 hours	$

3. If electric rates increase by 10%, the total amount per month for electric bills would be . . .

Seattle $_____

Minneapolis $_____

Chicago $_____

4. From the chart on page 146, you could conclude that:
   a) There are more electric power sources in the West, thus accounting for cheaper costs.
   b) More people use electricity in the East than in the West.
   c) Southern and northern states show little difference in electric power rates.
   d) Electric bills tend to be higher in the eastern cities than in the western cities.
   e) Electric bills tend to be higher in western cities than in eastern cities.

COST OF UTILITIES: PAYING TELEPHONE BILLS

One of the major costs in a budget is the telephone bill. While local calls can mount up, long distance calls can ruin your budget.

There are several guidelines you can follow to reduce your bill. Call at hours when there are reduced rates. This time is generally in the evenings. Also, if possible, make use of weekend rates, which are lower than on weekdays. Try to call station to station rather than calling person to person.

Rates within Pennsylvania

Dial-direct Sample rates from **Greater Pittsburgh** to:	Weekday full rate First minute	Each additional minute	Evening 35% discount First minute	Each additional minute	Night & Weekend 60% discount First minute	Each additional minute
Allentown	.46	.31	.30	.21	.18	.13
Erie	.43	.28	.28	.19	.17	.12
Harrisburg	.44	.29	.29	.19	.18	.12
Lancaster	.46	.31	.30	.21	.18	.13
Lebanon	.44	.29	.29	.19	.18	.12
Philadelphia	.46	.31	.30	.21	.18	.13
Reading	.46	.31	.30	.21	.18	.13
Scranton	.46	.31	.30	.21	.18	.13
State College	.43	.28	.28	.19	.17	.12
York	.44	.29	.29	.19	.18	.12

Calling Card Charges—Customer Dialed
$.30 plus dial-direct rate in effect for each minute of the call.

Operator Assistance Charges

Coin Calls—Paid:
$.35 plus dial-direct rate in effect for each minute of the call.
Collect and Third Number Billed:
$.60 plus dial-direct rate in effect for each minute of the call.
Time and Charges, Operator-Dialed Station, Operator-Dialed Calling Card and Hotel Guest Calls:
$.85 plus dial-direct rate in effect for each minute of the call.
Operator Person-to-Person
$2.25 plus dial-direct rate in effect for each minute of the call.

The Calling Card and Operator Assistance charges are not discounted.

Rates to other states

Dial-Direct

Dial-direct Sample rates from **Greater Pittsburgh** to:	Weekday full rate First minute	Each additional minute	Evening 40% discount First minute	Each additional minute	Night & Weekend 60% discount First minute	Each additional minute
Atlanta, GA	.62	.43	.37	.26	.24	.18
Atlantic City, NJ	.59	.42	.35	.26	.23	.17
Boston, MA	.62	.43	.37	.26	.24	.18
Chicago, IL	.59	.42	.35	.26	.23	.17
Des Moines, IA	.62	.43	.37	.26	.24	.18
Detroit, MI	.58	.39	.34	.24	.23	.16
Houston, TX	.64	.44	.38	.27	.25	.18
Los Angeles, CA	.74	.49	.44	.30	.29	.20
New York, NY	.59	.42	.35	.26	.23	.17
St. Louis, MO	.62	.43	.37	.26	.24	.18
Seattle, WA	.74	.49	.44	.30	.29	.20
Washington, DC	.58	.39	.34	.24	.23	.16

Calling Card Charges—Customer Dialed

Miles	Rate	Plus
1-10	$ .60	Dial direct mileage rate
11-22	.80	in effect for each minute
23 and beyond	1.05	of the call.

Operator Assistance Charges

Operator Station Third number billed, coin telephone, collect, time and charges, hotel guest calls and Operator-dialed Calling Card Calls.

Miles	Rate	Plus
1-10	$ .75	Dial direct mileage rate
11-22	1.10	in effect for each minute
23 and beyond	1.55	of the call.

Operator Person-to-Person

Miles	Rate	Plus
All	$3.00	Dial direct mileage rate in effect for each minute of the call.

The Calling Card and Operator Assistance charges are not discounted.

1. You made 20 calls within the state of Pennsylvania last
    month, totaling $10.00. These calls were made on a
    weekday.
    How much money would you have saved if the same calls
    were made evenings or on weekends?

    evenings: $_____   weekends: $_____

2. Compute the costs of the following monthly telephone calls
    based on the rate schedule on the previous page.

Mr. Bill Telephone Pittsburgh, Pennsylvania	Account Number: 123-456		Date Payment Due: August 5, 1983	
CURRENT MONTH'S CHARGES (W=weekday; E=evening; NW=night and weekend)				
In State Calls (direct dial)	**Code**	**Total Time**	**Amount Charged**	**Totals**
Harrisburg	W	1 minute	$	
Erie	E	2 minutes	$	
York	E	3 minutes	$	
Allentown	W	2 minutes	$	
Harrisburg	W	1 minute	$	
			Total (In State Calls)	$
Outside the State (direct dial)				
Los Angeles	NW	3 minutes	$	
Houston	W	1 minute	$	
			Total (Outside the State)	$
Operator Assisted (person to person)				
Scranton	W	3 minutes	$	
Seattle	W	3 minutes	$	
			Total (Operator Assisted)	$
Local Telephone Service:	from 6/14 to 7/14		$	15.15
			TOTAL NOW DUE:	$

THE COST OF FOOD

One factor that has caused the rate of inflation to rise is the increase in food prices. Although breakfast is the most cost-effective meal, the price of putting bacon, eggs, and milk on the table has risen dramatically over the past decade.

Breakfast is said to be the most cost-effective meal because it is at this meal that families usually get the most for their money. There is less of a tendency to prepare too much food, so there is less waste and unused leftovers than at the other two meals.

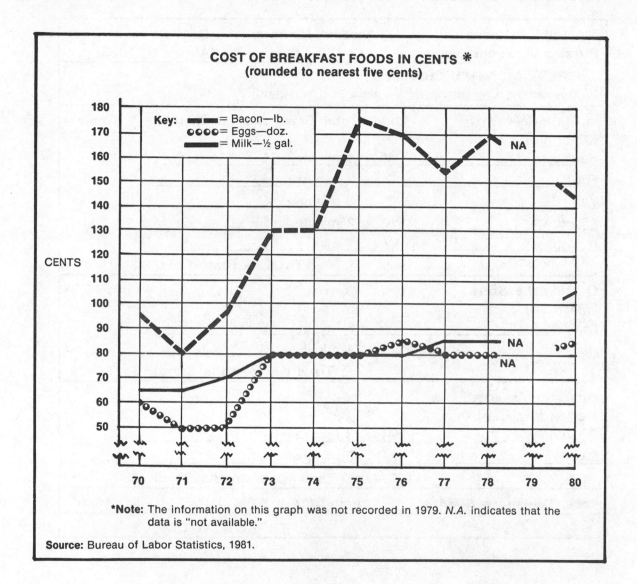

COST OF BREAKFAST FOODS IN CENTS ＊
(rounded to nearest five cents)

Key: ▬ ▬ = Bacon—lb.
○○○○ = Eggs—doz.
▬▬▬ = Milk—½ gal.

CENTS

＊Note: The information on this graph was not recorded in 1979. *N.A.* indicates that the data is "not available."

Source: Bureau of Labor Statistics, 1981.

1. Calculate the cost of these three breakfast foods for one week, assuming your family eats one pound of bacon, two dozen eggs, and drinks one gallon of milk.

   Figure the cost at 1970 prices and at 1980 prices:

1970		**1980**	
Bacon	$_____	Bacon	$_____
Eggs	$_____	Eggs	$_____
Milk	$_____	Milk	$_____
Total	$_____	**Total**	$_____

2. From problem 1 above, find the difference in breakfast food prices between one week in 1970 and one in 1980.

   $_____

3. From the graph on the previous page, determine which two food items have stayed about the same in price between 1973 and 1980.

   _____ and _____.

4. Compute the cost of these breakfast foods for a monthly budget (use figures for 1980), assuming there are 4 weeks in the month and your family eats the following:

   a) one pound of bacon per week
   b) one dozen eggs per week
   c) two gallons of milk per week

   Monthly Breakfast Foods: 1980

Bacon	$_____
Eggs	$_____
Milk	$_____
Juices	$____9.45____
Breads	$____4.20____
Cereals	$____6.00____
Total for month:	$_____

COSTS OF OPERATING A CAR

Most people now realize that the money that they pay to buy a car is only the beginning of their expenses. A buyer must also consider the cost of the daily wear and tear on the car, bills for maintaining and repairing the car, and the car's depreciation in value. These are factors in addition to monthly payments and the price of gas.

Depreciation is not as easily understood as a bill for repair or even the effect of wear and tear on a car. Depreciation is a term used to describe the decrease in dollar value of a car due to daily use. As a car gets older, it depreciates (decreases) in value.

The graph below shows the relative costs of operating a car. It also indicates that the number of miles that a car is driven per year tends to decrease over time. Why is this? One explanation is that today's cost-conscious drivers tend to "go easy" on an older car to guard against possible breakdowns.

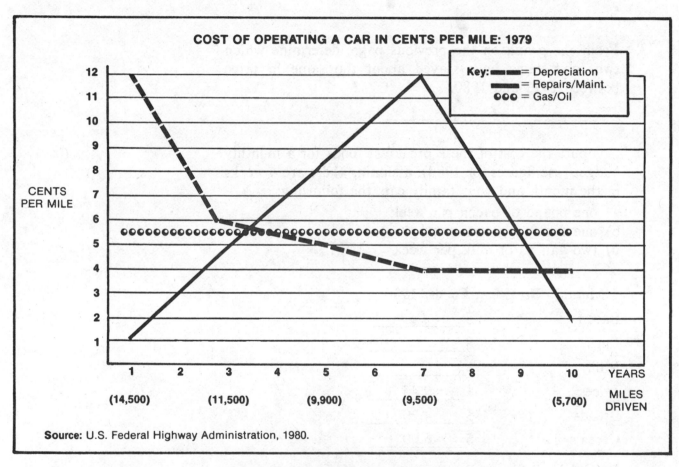

COST OF OPERATING A CAR IN CENTS PER MILE: 1979

Key: ■■■■ = Depreciation
■■■■ = Repairs/Maint.
⊙⊙⊙ = Gas/Oil

Source: U.S. Federal Highway Administration, 1980.

1. Compute the cost of depreciation, repairs and maintenance, gas, and oil in cents per mile for the first year of ownership.

   Depreciation _____ cents

   Repairs/Maintenance _____ cents

   Gas/Oil _____ cents

   Total _____ cents per mile for the first year.

2. If you drove the car 10,000 miles during the first year, the cost of operating the car would be $_____.
(Base your answer on Problem 1.)

3. During the _____ years, the cost of operating a car for the three categories is approximately the same.

    a) third and fourth
    b) fifth and sixth
    c) first and second
    d) sixth and seventh
    e) ninth and tenth

4. Determine the cost of operating your car during the seventh (7) year.

Depreciation   _____ cents

Repairs/Maintenance   _____ cents

Gas/Oil   _____ cents

Total   _____ cents per mile for year 7.

5. Compute the differences in operating costs between the first year (problem 1) and the seventh year (problem 4) in cents per mile:

a) Which year is the most costly? (Circle one.)   Year 1   Year 7

b) How much more costly? $_____ per mile.

c) Based on your judgment and the graph, answer the following questions:

    i) It is wise to sell a car after its seventh (7) year to avoid major repair bills.   True   False

    ii) You would need to keep a car for 7 or more years to see total operating costs begin to drop.   True   False

COSTS OF HOSPITALIZATION

The cost of medical care is expensive and on the rise. Often, the only protection against financial disaster is the purchase of a medical insurance policy. At first, monthly payments may seem an unnecessary cost, but the first major medical bill you face will seem much more manageable if you know that you will have help in paying it.

One of the main medical expenses you may face is the cost of hospitalization. The graph below compares the differences in hospital costs based on the size of the hospitals surveyed.

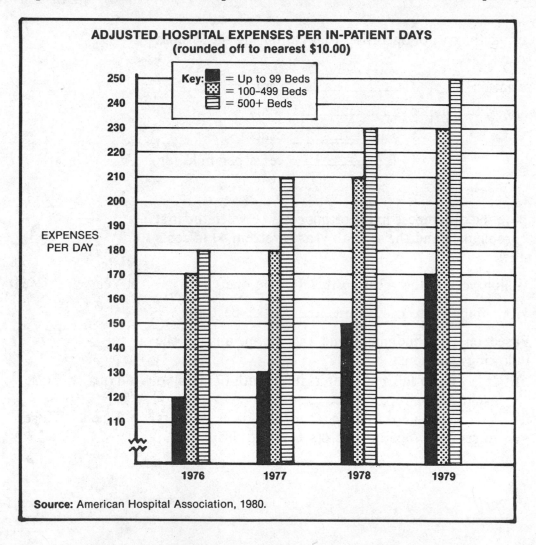

ADJUSTED HOSPITAL EXPENSES PER IN-PATIENT DAYS
(rounded off to nearest $10.00)

Key: ■ = Up to 99 Beds
▨ = 100–499 Beds
▥ = 500+ Beds

EXPENSES PER DAY

Source: American Hospital Association, 1980.

1. Determine the cost of a 10-day stay at each of the three hospitals below. Use 1979 figures.
    Hospital A: 55 Beds
    $_____
    Hospital B: 300 Beds
    $_____
    Hospital C: 650 Beds
    $_____

2. In 1978, you had knee surgery. The hospital you were in had 325 beds. Your stay was 8 days.

What was the total expense for this hospitalization?

$_____

Your health insurance policy paid $1,240. How much did you owe after the insurance paid its portion of the bill?

$_____

3. Find the difference between 1976 and 1979 hospital costs in a 500+ bed hospital.

$_____ per day

4. Answer True or False to the following statements, basing your answer on the graph:

   a) The larger the hospital, the greater the hospital expenses per day.   True   False

   b) No category of hospital increased costs more than $20 per day over a one year period.   True   False

   c) The year in which all categories of hospitals increased costs the same amount over the previous year was 1979.   True   False

CALCULATING DRIVING DISTANCES AND TIMES

Certain maps are designed to help the traveler estimate the distance in miles as well as the approximate time necessary to reach a specific place. On the map below, the numbers indicate the mileage between two cities. For example, the distance between Los Angeles and Phoenix is 389 miles.

Often, the time to drive distance is calculated by estimating the total number of miles it takes to travel in one hour. For example, if you travel 50 miles in one hour, then it will take you 8 hours to travel a total of 400 miles. (50 miles × 8 hours = 400 miles)

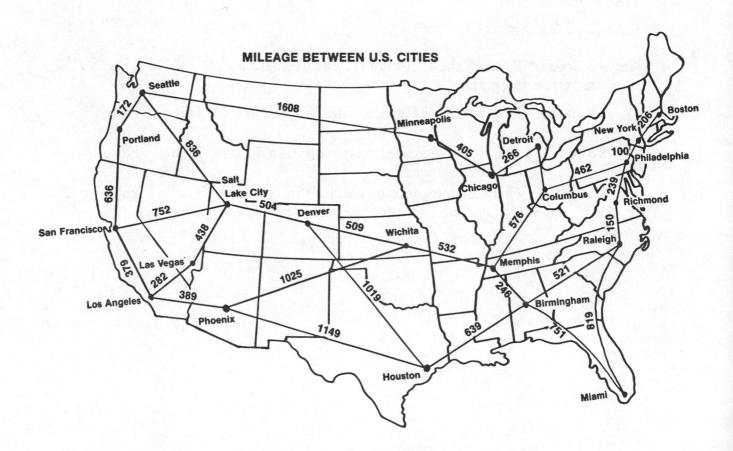

MILEAGE BETWEEN U.S. CITIES

1. Calculate the driving distances on the following trip (Although many routes are possible, use the most direct route.):

From	To	Distance in Miles
Seattle	San Francisco	_____
San Francisco	Los Angeles	_____
Los Angeles	Houston	_____
	Total Miles:	_____

2. Calculate the driving times for the following trips by dividing distance by miles per hour. (Round off to the nearest hour.)

From	To	Distance in miles	Time in hours (speed = 50 miles per hour)
Seattle	Minneapolis	_____	_____
San Francisco	Denver	_____	_____
Houston	Raleigh	_____	_____
Boston	Miami	_____	_____

3. You decide to take a trip from Phoenix to Wichita.

What is your gas mileage (in miles per gallon) if you used 50 gallons of gas (to the nearest tenth of a gallon)?

Note: $\dfrac{\text{total miles}}{\text{gallons used}}$ = miles per gallon

Answer: _____

What was your driving speed if it took 18 hours to drive this distance (round off your answer)?

Note: $\dfrac{\text{total miles}}{\text{hours driven}}$ = miles per hour (speed)

Answer: _____

4. According to the map, the shortest route is from _____ to San Francisco.

a) Houston to Phoenix to Los Angeles
b) Seattle to Salt Lake City to Los Angeles
c) Wichita to Salt Lake City
d) Minneapolis to Seattle
e) Denver to Salt Lake City to Los Angeles

READING A TOPOGRAPHICAL MAP

Hikers, backpackers, and forest rangers are some of the many people who are well acquainted with a topographical map. A topographical map shows the natural features of an area by means of lines and symbols. The lines on the map are called contour lines.

WILLAMETTE NATIONAL FOREST, OREGON

Source: U.S. Department of Agriculture, Forest Service, Pacific Northwest Region.

Each contour line represents a specific elevation. The next line over represents a change in elevation, usually 20 feet. By looking carefully at the map, you can tell if the elevation increases or decreases from one contour line to the next. You can see that steep terrain is represented by closely spaced contour lines, while flat terrain is represented by widely spaced lines."

Before reading the topographical map, make sure you are familiar with the following diagrams and symbols:

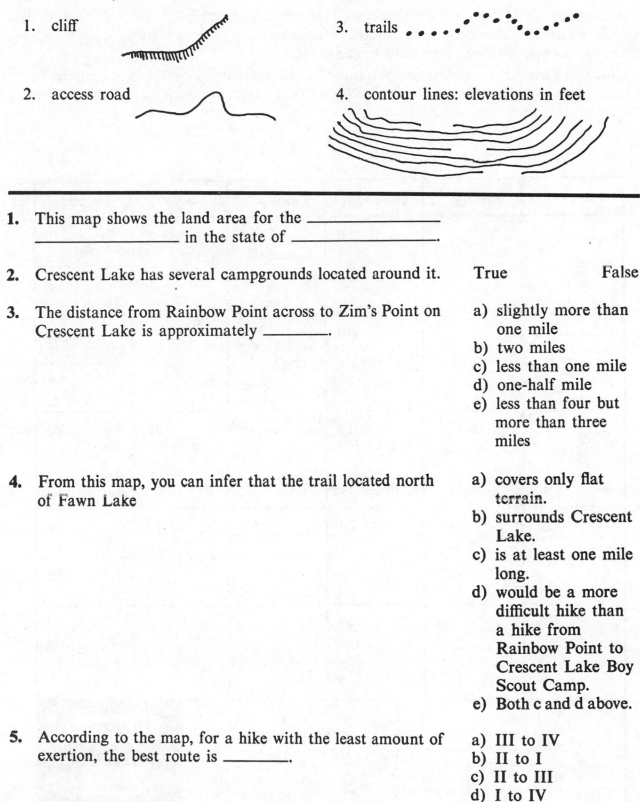

1. cliff

3. trails

2. access road

4. contour lines: elevations in feet

---

1. This map shows the land area for the _____ _____ in the state of _____.

2. Crescent Lake has several campgrounds located around it.   True   False

3. The distance from Rainbow Point across to Zim's Point on Crescent Lake is approximately _____.

   a) slightly more than one mile
   b) two miles
   c) less than one mile
   d) one-half mile
   e) less than four but more than three miles

4. From this map, you can infer that the trail located north of Fawn Lake

   a) covers only flat terrain.
   b) surrounds Crescent Lake.
   c) is at least one mile long.
   d) would be a more difficult hike than a hike from Rainbow Point to Crescent Lake Boy Scout Camp.
   e) Both c and d above.

5. According to the map, for a hike with the least amount of exertion, the best route is _____.

   a) III to IV
   b) II to I
   c) II to III
   d) I to IV
   e) IV to I

READING A CALENDAR

The calendar is a form of a chart that is used for scheduling activities from week to week and month to month. Calendars are important tools for keeping records of appointments, setting dates for events, making plans, and meeting goals.

Calendars may differ in their appearances but they generally contain the same features: a breakdown of each month by weeks and a view of the current month along with the months immediately before and after.

MONDAY	TUESDAY	WEDNESDAY	THURSDAY	FRIDAY	SAT./SUN.
1	2	3	4	5	Sat. 6
					Sun. 7
8	9	10	11	12	Sat. 13
					Sun. 14
15	16	17 St. Patrick's Day	18	19	Sat. 20
					Sun. 21
22	23	24	25	26	Sat. 27
					Sun. 28
29	30	31			

February						
S	M	T	W	T	F	S
		1	2	3	4	5	6
7	8	9	10	11	12	13
14	15	16	17	18	19	20
21	22	23	24	25	26	27
28						

April						
S	M	T	W	T	F	S
					1	2	3
4	5	6	7	8	9	10
11	12	13	14	15	16	17
18	19	20	21	22	23	24
25	26	27	28	29	30	

MARCH

1983

1. The calendar shown on the previous page represents the month of _____ in the year _____.

2. March 17 is _____ Day.

3. March consists of exactly four complete weeks.   True   False

4. The month of March follows April.   True   False

5. Perform the necessary scheduling as indicated below. Place all activities on the calendar shown on the previous page:

   a) Call and schedule an appointment for your boss, Mr. Parks, with Mrs. Ross, the tax consultant, on the first Friday in March, at 10:30 a.m.

   b) You arranged an airplane flight for Mr. Parks who will be in New York City for the last three days of the month.

   c) Mr. Williams called to set up an appointment to see Mr. Parks on Wednesday, March 10, at 10:30 a.m.

   d) Mrs. Ross called to cancel her previous appointment and asked to reschedule on Wednesday, March 24, at 9:00 a.m.

   e) Mr. Parks will review employees' files on the last Thursday of the month.

   f) Schedule two board meetings: one on the 9th of March; the other meeting is to be held 10 working days after the previous board meeting. Both meetings are to start at 1:00 p.m.

   g) Notify Mr. Parks five working days before the end of the month to prepare payroll checks.

ANSWER KEY

Graph Skills Inventory, Pages 1–4

1.	Japan	13.	Police officers
2.	6 million	14.	$21,000
3.	True	15.	False
4.	False	16.	True
5.	c	17.	b
6.	b	18.	b
7.	widowed; single	19.	food stamps monthly
8.	8.7	20.	21
9.	True	21.	False
10.	False	22.	True
11.	a	23.	c
12.	c	24.	b

Pages 12–13

1. a) Average Hourly Earnings In Different Job Categories
   b) Services; Wholesale and Retail Trade; Construction; Manufacturing; Transportation & Public Utilities; Mining; Finance, Insurance & Real Estate

2.	$2.00	7.	b
3.	Construction	8.	c
4.	False	9.	d
5.	True	10.	d
6.	False		

Pages 14–15

1. U.S. Department of Labor, Bureau of Labor Statistics

2.	1969; 1981	7.	a
3.	percent	8.	c
4.	False	9.	a
5.	True	10.	c
6.	False		

Pages 16–17

1.	projected	6.	False
2.	10 million	7.	b
3.	women and men	8.	c
4.	True	9.	b
5.	False	10.	d

Pages 18–19

1.	skilled jobs	6.	c
2.	300,000	7.	b
3.	True	8.	a
4.	False	9.	e
5.	True	10.	d

Pages 20–21

1.	Maximum Unemployment Benefits		
2.	Florida	7.	b
3.	True	8.	e
4.	True	9.	b
5.	False	10.	a
6.	c		

Pages 24–25

1. a) The Federal Budget Dollar: Where It Goes
   b) Grants; Net Interest on Loans; Benefit Payments; National Defense; Other Federal Operations

2.	individuals	7.	c
3.	national defense	8.	e
4.	False	9.	d
5.	True	10.	c
6.	False		

Pages 26–27

1.	$875,689	6.	True
2.	Alcoholism & Drug Abuse	7.	d
		8.	c
3.	10	9.	b
4.	False	10.	d
5.	False		

Pages 28–29

1.	estimated	5.	False
2.	322; 292	6.	True
3.	Housing, Health; Defense; Social Services	7.	a
		8.	b
		9.	d
4.	True	10.	b

Pages 30–31

1.	monthly income	6.	d
2.	19%	7.	b
3.	False	8.	a
4.	True	9.	c
5.	False	10.	a

Pages 32–33

1.	Officers; Enlisted	6.	b
2.	c	7.	b
3.	False	8.	d
4.	True	9.	d
5.	False	10.	d

Pages 36–37

1. a) Average weight for women 5′ 7″ tall
   b) average weight in lbs.
   c) age groups
2. 20–24   7. c
3. 158   8. c
4. True   9. d
5. False   10. c
6. False

Pages 38–39

1. average weight   6. True
2. 6′ tall   7. e
3. 20; 69   8. a
4. False   9. e
5. True   10. b

Pages 40–41

1. men; women   6. True
2. 20–24; 70–79   7. d
3. 160   8. a
4. True   9. d
5. False   10. d

Pages 42–43

1. third party; private consumers
2. 1975   7. b
3. True   8. c
4. True   9. c
5. True   10. b
6. d

Pages 44–45

1. fatally injured   6. b
2. 40   7. e
3. True   8. a
4. False   9. d
5. d

Pages 48–49

1. quarters of a year   6. False
2. cents per pound   7. b
3. 4   8. b
4. False   9. c
5. True   10. b

Pages 50–51

1. leaded premium   6. True
2. cents; gallon   7. b
3. 7   8. d
4. False   9. e
5. False   10. e

Pages 52–53

1. clothing, medical care, food/beverage
2. 1967   7. b
3. medical care   8. b
4. True   9. e
5. True   10. a
6. False

Pages 54–55

1. 18   6. d
2. 8%   7. b
3. True   8. d
4. False   9. e
5. True   10. a

Pages 56–57

1. electricity, telephone, gas
2. electricity; telephone
3. False   7. b
4. False   8. a
5. True   9. c
6. e   10. e

Pages 58–65

1. education   17. United States; 3
2. yearly income   18. 230 million
3. True   19. False
4. False   20. True
5. a   21. c
6. b   22. b
7. b   23. c
8. e   24. c
9. death; U.S.   25. Beef; Pork
10. 9.6%   26. 1979; 65
11. False   27. False
12. True   28. False
13. a   29. c
14. c   30. d
15. d   31. c
16. e   32. b

Pages 66–69

1. Central; Mountain   7. miles; weight
2. time   8. 1; 19 lbs. 15 oz.
3. False   9. True
4. True   10. False
5. e   11. b
6. c   12. d

ANSWER KEY (continued)

Pages 76–77

1. extraction; upper denture; bridge
2. dental services   7. a
3. 12 city   8. d
4. False   9. c
5. True   10. b
6. True

Pages 78–79

1. Johnson Community College
2. Mon., Jan. 26; Sat., Jan. 31
3. 787–1984   7. b
4. True   8. d
5. False   9. c
6. False   10. b

Pages 80–81

1. Suisan-Fairfield; Martinez; Richmond
2. 15; 18
3. sleeping car service
4. True   8. d
5. False   9. b
6. True   10. b
7. d

Pages 82–83

1. actual   6. d
2. tax preparation   7. c
3. True   8. c
4. True   9. b
5. False   10. c

Pages 84–85

1. 8,000; 9,000
2. single; married and filing jointly; married and filing separately; head of household
3. True   7. a
4. False   8. b
5. True   9. e
6. e

Pages 86–91

1. mileage   12. False
2. miles; cities   13. c
3. True   14. d
4. False   15. c
5. a   16. b
6. c   17. car; 13
7. e   18. 1000; 5000
8. b   19. True
9. depart   20. True
10. Portland;   21. b
    Seattle;   22. c
    San Francisco;   23. d
    Los Angeles   24. c
11. False

Pages 92–95

1. mountain range   10. c
2. latitude   11. d
3. False   12. 600
4. True   13. rainfall; inches
5. c   14. 40-59
6. downtown   15. True
7. east   16. False
8. False   17. d
9. True   18. d

Pages 98–99

1. northwest   4. southwest
2. southeast   5. south central
3. northeast   6. southeast

Pages 100–101

1. longitude   4. 30
2. latitude   5. south
3. equator

Pages 102–103

1. C-5; G-7; E-4   5. Portugal
2. France   6. True
3. Baltic Sea   7. False
4. Hamburg   8. False

Pages 104–105

1. a) 900   b) 400   3. D-5; 975; southeast
    c) 450   d) 300   4. A-1; 900; northwest
2. 200

Pages 110–111

1.	Arctic Circle	6.	True
2.	1000 miles	7.	False
3.	U.S.; Canada	8.	c
4.	80 degrees	9.	c
5.	True	10.	e

Pages 112–113

1.	Guatemala; Honduras		
2.	Guatemala; Belize		
3.	Cuba	7.	b
4.	east	8.	c
5.	True	9.	e
6.	b	10.	d

Pages 116–117

1.	San Francisco	6.	True
2.	A-4	7.	True
3.	east	8.	c
4.	False	9.	b
5.	False	10.	e

Pages 118–119

1.	B-3	6.	False
2.	Chicago	7.	c
3.	Lake Michigan	8.	a
4.	False	9.	e
5.	True	10.	c

Pages 122–123

1. military manpower; population
2. six
3. January 1980
4. 1 million–4,999,999; 5000–9999
5. False
6. True
7. True
8. b
9. d
10. c

Pages 124–125

1.	home heating; 5	6.	a
2.	electricity	7.	a
3.	A	8.	b
4.	False	9.	e
5.	True	10.	d

Pages 126–131

1.	300	13.	d
2.	Mekong	14.	b
3.	False	15.	c
4.	True	16.	b
5.	b	17.	unemployment; 1980
6.	d	18.	8% and over
7.	a	19.	True
8.	e	20.	False
9.	5.3	21.	b
10.	A-7	22.	c
11.	False	23.	b
12.	True	24.	b

Pages 132–140

1.	General Motors Corp.; 200,000	19.	e
		20.	d
2.	False	21.	Agriculture
3.	d	22.	False
4.	d	23.	a
5.	unemployed	24.	c
6.	True	25.	Colorado; Arizona
7.	b	26.	True
8.	c	27.	e
9.	18	28.	c
10.	False	29.	G-3
11.	c	30.	True
12.	d	31.	b
13.	local	32.	c
14.	True	33.	aid to dependent children
15.	b		
16.	b	34.	False
17.	85A; North Central	35.	d
18.	False	36.	a

Pages 142–143

1. True
2. False
3. c
4. b
5. Over budget

Income:
Take-Home Pay = $1,000.00

Expenses:		Adjustments (if any)	Corrected Budget
Food	$ 210	0	210
Clothing	100	−20	80
Personal Care	18	−3	15
Housing	300	0	300
Medical Care	100	0	100
Transportation	145	−5	140
Recreation	130	−60	70
Savings	55	0	55
Other	40	−10	30
Total Expenses:	$1098	−$98	$1000

ANSWER KEY (continued)

Pages 144–145

1. $1680
2. Atlanta
3. d
4. c
5. yes; no; yes; no; yes

Pages 146–147

1. $100
2. $35; $50; $110; $200; $400
3. $33; $77; $121
4. d

Pages 148–149

1. $3.50; $6.00
2.

Mr. Bill Telephone Pittsburgh, Pennsylvania	Account Number: 123-456	Date Payment Due: August 5, 1983
CURRENT MONTH'S CHARGES (W=weekday; E=evening; NW=night and weekend)		

In State Calls (direct dial)	Code	Total Time	Amount Charged	Totals
Harrisburg	W	1 minute	$ .44	
Erie	E	2 minutes	$ .47	
York	E	3 minutes	$ .67	
Allentown	W	2 minutes	$ .77	
Harrisburg	W	1 minute	$ .44	
			Total (In State Calls)	$ 2.79

Outside the State (direct dial)				
Los Angeles	NW	3 minutes	$ .69	
Houston	W	1 minute	$ .64	
			Total (Outside the State)	$ 1.33

Operator Assisted (person to person)				
Scranton	W	3 minutes	$ 3.33	
Seattle	W	3 minutes	$ 4.72	
			Total (Operator Assisted)	$ 8.05

Local Telephone Service:	from 6/14 to 7/14	$ 15.15
	TOTAL NOW DUE:	$ 27.32

Pages 150–151

1.

	1970		1980
Bacon	$ .95	Bacon	$ 1.45
Eggs	$ 1.20	Eggs	$ 1.70
Milk	$ 1.30	Milk	$ 2.10
Total	$ 3.45	**Total**	$ 5.25

2. $1.80
3. eggs and milk
4. a) $5.80 b) $3.40 c) $16.80
   Month: $45.65

Pages 152–153

1.

Depreciation	12	cents
Repairs/Maintenance	1	cents
Gas/Oil	5½	cents
Total	18½	cents per mile for the first year.

2. $1,850
3. a
4.

Depreciation	4	cents
Repairs/Maintenance	12	cents
Gas/Oil	5½	cents
Total	21½	cents per mile for year 7.

5. a) year 7 b) .03 c) i: False ii: True

Pages 154–155

1. a) $1700 b) $2300 c) $2500
2. $1680; $440
3. $70
4. a) True b) False c) True

Pages 156–157

1.

From	To	Distance in Miles
Seattle	San Francisco	808
San Francisco	Los Angeles	379
Los Angeles	Houston	1538
	Total Miles:	2725

2.

From	To	Distance in miles	Time in hours (speed = 50 miles per hour)
Seattle	Minneapolis	1608	32
San Francisco	Denver	1256	25
Houston	Raleigh	1160	23
Boston	Miami	1514	30

3. 20.5 mpg; 57 mph
4. c

Pages 158–159

1. Willamette National Forest; Oregon
2. True   4. e
3. b   5. a

Pages 160–161

1. March; 1983   **3.** False
2. St. Patrick's Day   **4.** False
5.

MONDAY	TUESDAY	WEDNESDAY	THURSDAY	FRIDAY	SAT./SUN.
1	2	3	4	5 *Mrs. Ross @ 10:30*	Sat. 6
					Sun. 7
8	9 *Board Meeting @ 1:00*	10 *Mr. Williams @10:30*	11	12	Sat. 13
					Sun. 14
15	16	17 *St. Patrick's Day*	18	19	Sat. 20
					Sun. 21
22	23 *Board Meeting @ 1:00*	24 *Mrs. Ross @ 9:00*	25 *Employee Filed. Prepare payroll cks.*	26	Sat. 27
					Sun. 28
29 *Mr. Parks in new York*	30	31 →			

February						
S	M	T	W	T	F	S
		1	2	3	4	5	6
7	8	9	10	11	12	13
14	15	16	17	18	19	20
21	22	23	24	25	26	27
28						

April						
S	M	T	W	T	F	S
					1	2	3
4	5	6	7	8	9	10
11	12	13	14	15	16	17
18	19	20	21	22	23	24
25	26	27	28	29	30	

MARCH
1983